Storage and Scarcity

In an era of abundance, at least part of humanity has stopped thinking about the future provision of basic vital resources such water, energy and food. Storage actions, with all their variants whether real or imagined, are sources of innovation in the provision and treatment of crucial resources.

This book explores the notion of water, food, energy and biodiversity storage as a response to a new era of scarcity. It analyses a variety of examples of multilevel storage policies, consumers' practices and local organisations, such as the industry and practices of food conservation, the need to stock agricultural produce, the role of artificial water basins in controlling floods and droughts and the development of batteries able to compensate for the intermittence of renewable energy sources. Storage and self-sufficiency can be achieved in many technical ways, at different territorial levels and according to different policies or philosophies. Being more a grasshopper or an ant – the two extreme positions – depends not only on the technologies available but also on different analyses of the environment and different attitudes to the future.

This book offers an environmentalist perspective that uncovers hidden or absent activities of ultramodern societies that will be useful to students of environmental sociology as well as those researching and studying at the interface of environmental studies and geography.

Giorgio Osti is a rural and environmental sociologist at the University of Trieste, Italy.

Routledge Studies in Environmental Policy and Practice

Series Editor: Adrian McDonald, University of Leeds, UK

Based on the Avebury Studies in Green Research Series, this wide-ranging series still covers all aspects of research into environmental change and development. It will now focus primarily on environmental policy, management and implications (such as effects on agriculture, lifestyle, health etc.), and includes both innovative theoretical research and international practical case studies.

Also in the series

Communities in Transition: Protected Nature and Local People in Eastern and Central Europe
Saska Petrova
ISBN 978 1 4094 4850 1

Sustainability and Short-term Policies
Improving Governance in Spatial Policy Interventions
Edited by Stefan Sjöblom, Kjell Andersson, Terry Marsden and Sarah Skerratt
ISBN 978 1 4094 4677 4

Energy Access, Poverty, and Development
The Governance of Small-Scale Renewable Energy in Developing Asia
Benjamin K. Sovacool and Ira Martina Drupady
ISBN 978 1 4094 4113 7

Tropical Wetland Management
The South-American Pantanal and the International Experience
Edited by Antonio Augusto Rossotto Ioris
ISBN 978 1 4094 1878 8

Rethinking Climate Change Research
Clean Technology, Culture and Communication
Edited by Pernille Almlund, Per Homann Jespersen and Søren Riis
ISBN 978 1 4094 2866 4

A New Agenda for Sustainability
Edited by Kurt Aagaard Nielsen, Bo Elling, Maria Figueroa and Erling Jelsøe
ISBN 978 0 7546 7976 9

Storage and Scarcity
New practices for food, energy and water

Giorgio Osti

LONDON AND NEW YORK

First published 2016 by Routledge

2 Park Square, Milton Park, Abingdon, Oxfordshire OX14 4RN

52 Vanderbilt Avenue, New York, NY 10017

Routledge is an imprint of the Taylor & Francis Group, an informa business

First issued in paperback 2020

Copyright © 2016 Giorgio Osti

The right of Giorgio Osti to be identified as author of this work has been asserted by him in accordance with sections 77 and 78 of the Copyright, Designs and Patents Act 1988.

All rights reserved. No part of this book may be reprinted or reproduced or utilised in any form or by any electronic, mechanical, or other means, now known or hereafter invented, including photocopying and recording, or in any information storage or retrieval system, without permission in writing from the publishers.

Notice:
Product or corporate names may be trademarks or registered trademarks, and are used only for identification and explanation without intent to infringe.

Library of Congress Cataloging in Publication Data
Names: Osti, Giorgio, author.
Title: Storage and scarcity : new practices for food, energy and water / by Giorgio Osti.
Description: Farnham, Surrey, UK ; Burlington, VT : Routledge, 2016. | Series: Ashgate studies in environmental policy and practice | Includes bibliographical references and index.
Identifiers: LCCN 2015043717| ISBN 9781472483010 (hardback : alk. paper) | ISBN 9781472483034 (epub)
Subjects: LCSH: Food security. | Energy security. | Water security. | Food—Storage. | Energy storage. | Water—Storage. | Environmental responsibility.
Classification: LCC HD9000.5 .O85 2016 | DDC 333.71/6—dc23
LC record available at http://lccn.loc.gov/2015043717

ISBN: 978-1-4724-8301-0 (hbk)
ISBN: 978-0-367-66830-3 (pbk)

Typeset in Times New Roman
by Swales & Willis Ltd, Exeter, Devon, UK

To Bruno, a friend forever

Contents

List of illustrations ix
Preface xi
Acknowledgements xvii

1 **Accumulation versus networking: the terms of the storage issue** 1

   1.1 *Storage, an eclipsed issue 1*
   1.2 *Between storage and security 3*
   1.3 *Why storage is once again important 7*
   1.4 *Social analysis of storage 11*

2 **The modularization of food processing and consumption** 19

   2.1 *Food conservation, a wide-ranging and ambivalent issue 19*
   2.2 *Food security practices 25*
   2.3 *Healthy food: storage as preservation 34*
   2.4 *Organizations opposed to food waste 40*
   2.5 *Rituals of food storage 47*
   2.6 *Rationing and rationalizing 56*

3 **Water storage: a multidimensional task** 58

   3.1 *The long march toward water channelling 58*
   3.2 *River banks, detention basins and floodable areas 65*
   3.3 *The art of water conservation and harvesting 77*
   3.4 *The big task of small farm ponds 87*
   3.5 *In between network and storage 93*

4 **Energy storage** 97

   4.1 *The issue 97*
   4.2 *The legal, technological and social frames 103*

- 4.3 *The Tartar Steppe of energy storage 108*
- 4.4 *Power storage systems embedded in national policies 114*
- 4.5 *Thermal energy, the Cinderella of storage 126*
- 4.6 *Conclusions 135*

5 Long-term life storage   137

- 5.1 *Nature reserves, botanical gardens and seed banks 139*
- 5.2 *Beyond land sparing and land sharing 152*
- 5.3 *The mutual exchanges of biodiversity and storage 157*
- 5.4 *Mosaics of biodiversity 164*

6 Multi-storey: the fortune of the grasshoppers and the ants   167

- 6.1 *Policies for storage: lobbies and the need for space 172*

References   176
Index   200

Illustrations

Boxes

1.1	The economic theory of storage	5
2.1	Cooperative Terremerse, Ravenna, Italy	32
5.1	The Trentino bear destined for an animal sanctuary?	153

Figures

0.1	Inclusion of storage practices in a set of topics according to a circular causation model	xiii
1.1	Polysemy of storage practices	7
1.2a	Vicious circle of consumption	10
1.2b	Virtuous circle of consumption	10
1.3	The network/storage contraposition according to environmental and social drivers	18
2.1	Classification of the main forms of food storage according to the intersection of (self-) control versus agency and value rationality versus means-end rationality	25
2.2	Circularity between development of food storage techniques and safety measures	40
2.3	Main cultural meanings of food storage	52
2.4	Traditional practices of local food reserves in Asian culture	53
2.5	Percentage of International Network of Preppers website visits according to the country of the visitor	54
2.6	Locations of preparedness practices along the civil protection–survivalism continuum	55
3.1	Evolution of potable water provision systems: points of departure, model adopted and outcomes according to the importance of types of storage	85
4.1	Rated MW power of US grid storage projects, including announced projects	103
4.2	Maturity of energy storage technologies	107
4.3	Reasons for low penetration of energy storage	110

4.4	Graphic representation of the storage systems divide	120
4.5	Continuum of storage policy intensity	122
5.1	Classification of biodiversity storage forms according to the in situ–ex situ continuum	138
5.2	Conceptual model of the continuum of scales at which biodiversity conservation and agriculture can be integrated	155
5.3	The honeycomb of biodiversity: socio-cognitive dimensions of biological diversity	156

Tables

2.1	Example of energy (kcal) used to store 455 g of corn	43
3.1	Diagram of the three patterns of water security according to social, political and economic aspects	72
3.2	Relationship between types of water connection and water domains	93
4.1	Classification of energy storage methods	98
4.2	Evolution of energy storage systems according to their periods of major expansion	99
4.3a	Crucial factors for energy storage systems in four areas	116
4.3b	Crucial factors for energy storage systems in four areas (presence of a movement)	117
4.4	Forms of thermal energy storage according to their sources, methods and uses	128
5.1	A selection of seed banks, their year of constitution and institutional profile	146
5.2	Main features of three forms of biodiversity storage	150
6.1	Features of storage practices in terms of temporal continuity and module integration within networks	168

Preface

In the era of abundance, at least part of humanity has stopped thinking about the future provision of basic vital resources: food is easily found in convenient and nearby stores; water and energy are delivered directly to the home through efficient and ramified grids. Distribution has progressed to such an extent that almost all goods can be delivered to the home through e-commerce and courier services. If anything, the problem is low income, and the consequent inability to afford the costs of such an abundant availability of goods on our own doorsteps. In developing countries and some remote areas of developed countries this abundant provision is still lacking. This is a problem not only of widespread poverty, but also of bad distribution. Energy or water grids need large initial public investment and a secure number of users. Distribution of packaged goods, such as food, has less need of a large preliminary investment, but it too requires a sufficient number of nearby customers.

Fresh meat and vegetables can be bought every morning at the market; drinkable water and calorific gas constantly arrive through pipelines. Very rarely is there a shortage or a break in provision. Energy companies usually provide good maintenance of services: a blackout may interrupt the electricity supply for some minutes, hours, or even days, but confidence that supply will resume is high, and confirmed by regular past provision. Again, in many parts of the world there is no such regular provision of vital resources. Water and electricity arrive for a few hours a day, and their quality is poor. A large proportion of humanity has no electricity connection at all. In both situations – stable and precarious provision – the objective is the same: to create a regular and ramified grid with abundant consumption, which is a political aim, or at least an aspiration. In any case, it is a model, a target, a benchmark with which to compare less fortunate situations.

But the era of abundance may be coming to an end (Barbier 2011; Brown 2012; Manders 2011). Basic resources are increasingly insufficient, for a number of reasons: intractable waste, population growth and the depletion of deposits. The problem is well known and sufficiently proven. What remains obscure is whether efficiency measures are enough, or whether a radical redistribution of goods among social classes and areas of the planet is necessary. Fiscal policies – like carbon taxes or progressive taxation – are insufficient because they negatively affect two crucial aspects of the new era of scarcity:

people's lifestyles and goods distribution systems. The former are linked to imponderable cultural variables, the latter are embedded in socio-technical networks that have arisen in the period of abundance and are resistant to any change that will reduce their provision capacity. Storage enters into both these issues: it concerns lifestyles, and it challenges the grid systems of distribution.

The extent to which authorities and communities are concerned about these two issues is not at all clear. Appeals to reduce consumption are very frequent (Humphery 2009), if nothing else because there are serious secondary effects (for example, obesity). There is less awareness of distribution methods, in particular as regards storage and provision. Before the era of abundance, storage and provision of water, food and energy were major problems that were addressed in many ways. The need for food conservation generated a wide range of techniques and habits (salted food, seasonal diets); water was collected and kept in rooftop tanks; energy was regularly provided thanks to animals and serfs maintained in passable conditions. Of course, some techniques – typically refrigeration and grid distribution – have reduced the problem of storage; but because they are energy-demanding, the problem of shortage arises again. It is better to process agriculture goods immediately, in order to avoid the costs of long-term conservation, but new risks ensue from the use of artificial preservatives. Electrical energy is probably the best example of the dilemma of modern technologies: once electricity has been produced, it is almost impossible to store, and must be consumed.

For various reasons, therefore, provision and storage systems should be placed at the centre of our attention. They require action at three levels: the household, the local community and the nation. The question is whether each level is organizing new forms of basic goods provision and storage that are able to reduce the depletion of resources and prevent shortage crises. The social analysis of these levels considers practices, policies and long-term structures like beliefs, cultures and institutions. In an uncertain scenario on the role of storage in preventing food, energy and water crises, an expressive divide is between *ants* and *grasshoppers* (in Romance languages, called *cicadas*). Some households, communities and countries are more zealous than others in organizing less costly systems for the provision and storage of vital resources. There is also a dualism in terms of stocks and flows: almost a philosophical dilemma between stubborn self-sufficiency and bonding relationships.

This philosophical dilemma has a political consequence as well. In periods of abundance, the free market allocates goods better: people are more confident in the invisible hand's ability to provide everyone with the best chances of finding a commodity, a job, a dwelling. In periods of scarcity, people are less confident in automatic mechanisms: they distrust each other, and start hoarding. External forces of regulation must intervene to calm people and distribute goods. Of course, in these situations the abuse of force is likely. In the context of penury or disasters, some communities concentrate more on prevention, storing living goods and training for civil protection.

To conclude: prevention, storage and self-sufficiency can be achieved in many technical ways, at different territorial levels, and according to different policies or philosophies. Being more a grasshopper or an ant – the two extreme positions – depends not only on the technologies available, but also on different analyses of the environment and different attitudes to the future. Hence, faith and religion are important factors linked to diverse cosmologies and eschatologies. Provision and storage systems are thus at the centre of a wide field of analysis to be explored in light of the end of the era of abundance. This is an environmentalist perspective that promises to uncover hidden or absent activities of ultramodern societies.

Most of the literature on vital resource storage varies from the scenarios analysis – with a range from complacency to catastrophism – and detailed suggestions on what to do easily drift into normative attitudes, plain advice and ingenuities. The bulk of this book lies somewhere in between: it analyses and reports many cases of family-, community- or regional-level responses to emergencies (preparedness) that already exist and are underestimated. These cases are more instructive than general analyses because they are practical, visible, easy to handle and offer a starting point for gradual improvements. In fact, it is preferable to move little by little toward a more sustainable society than to wait for an ecological re-foundation (see Foster 2010).

An example will illustrate the primacy of practices. Fresh water management will be studied at all scales of analysis, and a plurality of public and private actors considered (Norman, Cook and Cohen 2015). Those scales are interconnected by daily practices of administrators, technicians and water users' movements. It is fundamental to affirm that flood prevention has to be treated at the watershed level, but innovation occurs when a major detention basin is connected with the seasonal restoration of ditches made by land reclamation consortia and by each landowner, without forgetting the wastewater catchment basin cleaning of urban dwellers. The same goes for food, energy and biodiversity: multi-level policies have to be matched with grassroots practices.

The background literature for studying storage practices can be represented by Figure 0.1.

Increased pressures on ecosystems have provoked a depletion of vital resources, a decrease of biodiversity, and the introduction of dangerous substances in human

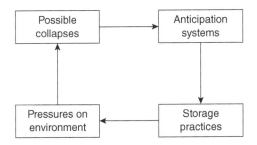

Figure 0.1 Inclusion of storage practices in a set of topics according to a circular causation model

and natural environments, with high production of intractable wastes. The *Driving Forces–Pressures–State–Impacts–Responses* (DPSIR) framework, despite its limitations (Maxim, Spangenberg and O'Connor, 2009; Carr et al. 2007; Svarstad et al. 2007), is a starting point frequently used by environmental analysts (Ness, Anderberg and Olsson 2010). The systemic nature of the DPSIR framework has been the basis for well-known predictions of the collapse of the world ecosystem since the celebrated Club of Rome study (Diamond 2005; Tainter 1990; Douthwaite and Fallon 2010; Perna 2011). The world is discovered to be more vulnerable, ecosystems less resilient, and climate events more bizarre (Waughray 2011). In this uncertain environment, some individuals (Paek et al. 2010) and some communities (Comfort 2005; Comfort et al. 2001) show the will and the capacity to pre-organize measures for dealing with possible adverse events. They develop anticipation (Poli 2010) or learning (Wenger 1998) systems that take the forms of public programmes, dedicated organizations, targeted devices, voluntary corps, daily habits: in a word, *practices* (Spaargaren 2011). Storage practices traverse all these social forms. Their novelty consists of using old forms of action typical of pre-industrial society rediscovered to face a new era of scarcity, new possible disasters, new environmental problems and – why not? – new ways to improve comfort and welfare. Often, storage is a *retro-innovation* (Stuiver 2006); but, like every human activity, it is a mix of rationality and rituals (Wynne 2013) with a gradient from the single individual attitude to the most complex organizations (Goffman 1983).

This book does not propose storage as the solution to the environmental crisis, but rather as a perspective, a point of view, that begins from that aspect of human activity which involves prevention, preparedness, accumulation, saving and subtraction from immediate consumption. This, hopefully, is the added value of this book: it highlights a new perspective on socio-environmental studies. Indeed, storage practices constitute an immense field of research. The majority of cases will come from Europe and North America. This is not to exclude emerging countries and relatively poor ones; it is a methodological choice. Western countries represent a homogenous domain of research: they face the same problems of mature overconsumption, and they have the most advanced grid systems for the delivery of items of every kind. Ideas and examples of change can and must come from these countries. With tangible changes, Western countries can negotiate reasonable world limits on natural resource extraction. Some sporadic references (especially with regard to energy storage) will be made to Asian countries. Ideal-type polarizations will not be easy, even if North America is more involved in grassroots preparedness than Europe, which seems internally diversified: the northern countries are more concerned with climate change and adaptation/mitigation measures, while the southern ones are less so. Many examples come from Italy, not only because it is the author's own country, but also because its long civilization, linked with a recent entry into the industrial era, puts storage practices in an intriguing position. In ancient but turbulent countries, such practices are very old and almost neglected, but are now urgently called on for an environmental renaissance.

Preface xv

The goal of the book is to answer three questions:

1 Is storage a category that is able to highlight new or neglected social phenomena?
2 Is storage an ideal type with heuristic capacities?
3 Is storage really middle-range – intermediate between the shallow and deep views of environmental issues?

Thus, the analysis will begin with storage as a purely cognitive stratagem, but the final ambition is to draw attention to useful new ways to meet the challenge of creating a more prosperous and equal society (Jackson 2009).

Chapters

Chapter 1 illustrates four points: (1) the anthropological background of storage, and the fact that it has been an archetype of human organization; (2) storage, especially food conservation, was an important source of social security for centuries until the industrial revolution; (3) the reasons why the importance of storage declined during the modernization period when societies substituted it with the instant delivery of goods through efficient networks; (4) the ways in which social sciences can foster renewed attention to the crucial nature of goods storage.

Chapter 2 focuses on food storage; it starts with a general overview of storage in terms of *food security*. The topic alternates between needs for social control and practices of self-sufficiency. A second theme is *food safety*, because of the great efforts made in every age to preserve food from diseases caused by forced periods of storage. The third theme is the rational fight against *food waste* conducted by intelligent organizations along the chain from food production to recycling. Finally, the plural world of *food symbols* is explored, in that storage is also a quasi-religious ritual.

Chapter 3 on water storage is introduced by a history of land reclamation. For centuries the diktat has been that water must drain away from land as soon as possible. Land reclamation and water flow acceleration increased usable surfaces, but created new insecurities. Slowing down the water's speed and creating detention basins are now seen as new ways to prevent floods. Water storage is also crucial for household uses (harvesting) and agricultural irrigation (farm ponds). In both cases, a mix of micro-storage structures connected to a network emerges as a good solution to the problem of scarce or too abundant water.

Chapter 4 views energy storage as a socio-technical *network*. There is a special link with the rise of renewable energy sources; their intermittence requires the complementary use of storage devices. Furthermore, energy storage entails more room for the final consumer and modifies the relationship with the grid energy supplier, usually prominent on the demand side. At first glance, energy storage is a way to sever the link with a network. However, the most intriguing situation is where there is a mix of energy self-provision, storage and exchange with a grid.

Chapter 5 explains that biodiversity conservation is the necessary complement to food, energy and water storage. It is a means to maintain entire *ecosystems* rather than single natural resources. After illustrating the main forms of biodiversity storage – natural parks, botanical gardens/animal sanctuaries and seed banks – the chapter shows the mutual advantages between human-aimed conservation and biodiversity. A mosaic or landscape ecology is finally presented as an attempt to arrive at a synthesis.

Chapter 6 summarizes the book's results in two analytical dimensions: temporal *continuity* of storage practices and *integration* of single modules. The four storage fields – food, water, energy and biodiversity – are inserted into this framework. Finally, possible storage policies are recommended.

Acknowledgements

Many thanks to Giovanni Carrosio, Kris Van Koppen, Stewart Lockie, Arturo Lorenzoni, Antonio Massarutto, Luigi Pellizzoni and Simone Arnaldi for reading and revising parts of the book; I also thank the staff at Routledge, Amanda Buxton, Katy Crossan and Margaret Younger, for their convinced support during the book preparation and Rachel Singleton and Huw Jones of Swales & Willis for their editorial assistance; sincere gratitude to Adrian Belton for his patient revision of the manuscript's language. I'd like especially to mention Bertrand Sadin and Giovanni Osti for elaborating the figure artwork. The academic atmosphere of the Department of Political and Social Sciences of Trieste University in Italy has been of great help.

In this book I present part of the survey 'Sustainable practices of everyday life in the context of the crisis: toward the integration of work, consumption and participation', funded by the Ministero dell'Istruzione, dell'Università e della Ricerca's Progetti di Ricerca di Interesse Nazionale (MIUR-PRIN) in 2010–2011 and co-ordinated by Laura Bovone (Università Cattolica di Milano), in collaboration with the Universities of Milano (coordinator Luisa Leonini), Bologna (coordinator Roberta Paltrinieri), Trieste (coordinator Giorgio Osti), Molise (coordinator Guido Gili), Roma 'La Sapienza' (coordinator Antimo Farro) and Napoli Federico II (coordinator Antonella Spanò).

1 Accumulation versus networking
The terms of the storage issue

1.1 Storage, an eclipsed issue

Storage is a primordial activity. It is so ancient that it is a behaviour shared with those animal species that practise the collection of seasonal food. But storage is probably also a turning point in the history of humanity, because food and energy storage mark the capacity to create conditions of stability allowing child education and community intellectual activities beyond mere physical reproduction. The late Neolithic period was distinguished by a greater capacity to store water and food in artificial containers. Anthropologists, especially those involved in archaeology, highlight the importance of storage in the increase of societal complexity (Hendon 2000), even though they emphasize that, at the dawn of society, storage, for example of food, happened at the household level. Higher levels of storage were socially more problematic and uncertain (Bale 2012).

Thereafter, improved ability to store food, energy and water gave some communities an advantage over others: it created the conditions for a more stable and predictable environment and the development of devices useful for agriculture and for war. Good storage enabled communities protected by walls to resist enemy sieges. At the same time, it was a source of power not only between communities, but also within communities. The differential capacity to store vital resources generated asymmetries in exchanges, with some actors more able to ration the goods and then to keep sufficient reserves for long time (Halstead and O'Shea 1982). That gave rise to the phenomenon of *surplus*: in other words, the production of an amount of foodstuff that exceeded survival needs. That surplus was the basis for a differential accumulation of wealth and power within communities and among them. It is difficult to establish a distinct threshold between the satisfaction of survival needs and food surplus; however, that does not impede being able to see the extraordinary power generated by storage capacities.

Storage is a basic criterion for the analysis of human organization, an archetype of associated human beings. It has been less thematized because modernity, conceived as a centuries-old dominant culture, has sought some sort of emancipation from the urgency of storage. This has happened for three main reasons: the development of networks and exchanges, the discovery of freezing technology, and the abundance of underground resources like oil.

At a time when everything was made easily available by means of rapid transport, the storage of many items became obsolete and expensive. Markets, with their rapid and low-cost capacity to provide every kind of *article of trade*, were the antinomy mechanism of storage. There came a time in the history of industrial production when companies thought that they could work 'just in time', drastically reducing the need for warehouses (Cheng and Podolsky 1993). In ideal terms, the market and the network behind it can be seen as the opposite of storage. They can provide every good requested almost in real time; the rapidity and completeness of such provision are the features of a working market. Indeed, scholars of organized spaces tell us that it is impossible to eliminate the *friction of distance* in communication, especially for material goods (Ellegård and Vilhelmson 2004). It is impossible to provide the latter in real time. Moreover, rapid transport has a great impact on the environment, and high energy costs. Thus, the simultaneous capacity of the market is an ideal. Nevertheless, its opposition to planned and huge warehouses remains real.

A second reason for the disappearance of the storage issue was the extraordinary development of freezing machinery. Before the introduction of such technology, the provision of cold to conserve food and drugs during the warm season was very demanding. Large caves had to be dug underground in order to keep ice created and transported during the cold season. Application of the Carnot cycle – the theory at the basis of artificial refrigeration – was successful for a set of concomitant reasons: the development of artificial insulation materials and the availability of electric energy to supply compressors. Furthermore, it was useful for creating both cold and heat, even if the latter advantage was fully developed later. The capacity to create refrigerators of all sizes – from the cold store to the car fridge – and the modularity of machines which combined a freezer and a refrigerator led to ubiquitous application of this technology. Note that this was a technology that had minimal negative secondary effects (ozone-depleting gases, primarily) and was cost-effective. In conclusion, successful storage technology obscured the atavistic problem of storage. We know, moreover, that refrigerators wrought a revolution in the organization of households and consumption styles. The extreme capacity to store through cold did away with the need to cook every day and keep an entire room as a cellar.

The third reason for the eclipse of the storage issue was the great abundance of cheap energy, as has already emerged above: transport facilities and freezing technology took advantage of a great availability of energy. Coal, and later oil, brought about the miracle of providing a copious source without rivalry with other land uses. In the past, energy provided by wood and animals created competition with arable land designated for the nourishment of human beings. Underground energy resources changed this situation and created the basis of a great demographic increase, as actually happened. From our point of view, another feature of coal and oil was fruitful for the destiny of humanity: their capacity to be stored quite easily, either by modulating their extraction or keeping them in simple containers. For biomass, wind, falling water and animal-origin energy, such storage has traditionally been more problematic. The conservation of fossil energy sources, including natural gas, is easy – a feature which largely explains the success of automobiles powered by such fuels. Again, storage capacity became a reason for neglecting a key factor in the organization of human life. The taken-for-granted qualities and quantities of fossil fuels covered up the problems of their reserves and of keeping them available and their use efficient.

This brief historical reconstruction shows that storage as an *archetypical factor* of human organization has not disappeared. Rather, it has been clouded, because of the improved capacity of some societies to find temporary alternative solutions. Not by chance, less-developed societies are still contending with the age-old problem of food, energy and water storage. This observation highlights a sort of territorial divide: storage in the period of industrialization became a secondary problem for more developed countries, while it retained all its urgency in the less developed ones. But now – and this is the main point – it is becoming a problem for the entire world, facing a new era of scarcity. The use of the progressive present tense is revealing, both because the issue is not fully recognized, and because the balancing point between immediate consumption and storage is rather mobile. We must abandon the idea of a neat dichotomy between the two aspects: the direct use and the storage of water, energy and food are intertwined.

1.2 Between storage and security

Storage is the conservation of vital resources like food, energy and water in dedicated containers so that they can be consumed during periods when they are not easily available, 'an activity involving the placement of useful material resources in specific physical locations against future needs' (Hendon 2000: 42). Attention should be paid to the special spatial–temporal coordination performed by storage: *it consumes space in order to save time*. Effective storage action needs to maintain resources for basic features like performance capacity and safety over time. For example, a freezer is able to keep meat calories almost unchanged for a long period without hazards to the health and security of consumers, but it has its *ecological footprint*. Two corollaries ensue from this thinking. The first is the clear instrumental meaning of storage: it is a practical action aimed at making the provision of some goods easier and more stable over a period of time. The second corollary is that *security* comprises *safety* in the sense that storage enables access to a product over time without provoking harm for the final users. In other words, it is a regular provision of healthy or non-damaging goods.

A point not contained in Hendon's definition concerns *distribution*: who is entitled to access the stored goods? As stated above, storage is a material action with serious political consequences. The managers of storage facilities acquire a great deal of power. For this reason, storage is usually regulated by public authorities, or indeed directly managed by them (Barquín 2005: 260). The intense presence of the state since the *ancien régime* in storage activities is not only a matter of justice – to ensure that everybody has essential resources – but also a matter of social control. The assumption is thus negative: people left alone tend to overconsume or to hoard. The former activity creates waste; the latter a false state of penury. Because people are irrational, a central authority must intervene, especially in periods of turbulence (war, bad weather, riots and so on). It is now clearer why storage management is linked to security in terms not only of stable provision, but also of politics. Also clearer is its ideal-typical opposition to the market mechanism, which implies actors' freedom and rationality, and very limited government intervention (Ó Gráda and Chevet 2002). On the contrary, a central bonding political action proves to be necessary to keep the community alive and under control.

Not by chance, some storage activities are classified under *security policy*, a field more linked to the fight against social disorder and poverty. Storage in this sense has a *redistributive function*: it allows the forced collection of the good from all members of the community and its distribution to poorer ones. It works in a way very similar to the ideal-type 'redistribution' of Karl Polanyi (1944). In the case of food, the storage centre works more or less as a selective dispenser: the authorities maintain a reserve of food constantly ready for emergencies or peaks in demand. In the case of water, it typically takes the form of intervention with tankers when the grid is absent or contaminated. In fact, grids for water were built for security reasons: to allow the regular (and comfortable) delivery of drinkable water to a neighbourhood. Later, social security worries facilitated the service's access to more peripheral areas. It is interesting that the delivery of vital resources like water and energy through grids changed the intervention to ensure social security: it moved the problem to the efficient functioning of the grid and to the limited quota of people unable to pay the service pipe fee. Thus, by means of the grid, social help became more precise and at the same more bonding. As storage has been theoretically conceived as an alternative to the grid, it is noteworthy that the former assures more autonomy to poor people, while the latter centralizes social security actions and makes them more managerial.

Storage thus traverses many fields. From the simple meaning of assuring the regular provision of resources, it easily acquires political and social ones. But there is a further meaning. According to public goods theory, the provision of certain goods via the market produces sub-optimal results. Commons are a case in point. When individual rational actors can use a good without paying a price for it, because it is open-access, such actors tend to exploit it endlessly. They do not consider the reproduction time of the good. The water of a river provides an example. The landowner upstream is tempted to use as much water as possible to irrigate his field because water is free and nearby, disregarding farmers downstream. We know that solutions for this free rider problem are many and disputable (Ostrom and Ostrom 1977).

For our purposes here, it is important to show that storage is part of larger-scale solutions to problems concerning commons. A dam at the top of a river can store a great deal of water and offer security of provision for the summer. Of course, the dam needs many other matters to be agreed among downriver landowners and authorities. Stored water in any case has a greater economic value than the river water itself, because it can be used according to the desires of producers and consumers. In other words, the main economic actors find the storage system to be a more suitable solution for coordination of their objectives (Smith and Thomson 1991: 96), eventually limiting the *tragedy of commons*.

When a good does not face market failure and is easily marketable, economists use the theory of storage (see Box 1.1). With this theory they assume the functional possibility of storing commodities, and they try to explain the variation of prices, generally in the stock exchange markets, according to the rule of demand and supply. They acknowledge, starting with Keynes (1938), that speculation is possible on the demand side because of the exchange of *futures* (Geman and Smith 2013). Nonetheless, speculation is also possible on the supply side, when great storage capacity is used by monopolies or oligopolies to impose their own prices on demanders. Storage itself, therefore, is not a source of speculation; it facilitates speculation when combined with dominant positions in the market or when financial tools are not sufficiently ruled and monitored.

Box 1.1 The economic theory of storage

Commodities are categorised as storable or non-storable. Nonstorable commodities include those where storage methods exist but are prohibitively expensive (in particular, the case of electricity) and where the commodity is the provision of a service (as in the shipping industry). The vast majority of commodities are storable. They are stored for several reasons:

- As a buffer against uneven or seasonal supply, as in the case of agricultural commodities, which have been stored in silos as early as 10,000 years ago.
- As a buffer against uneven demand, as in the case of most energy commodities, which are typically used more in winter for heating, and midsummer for cooling.
- As a buffer against any other supply or logistical disruption, which would otherwise necessitate the expensive pause of an industrial process.
- In recent years, for investment purposes within physically backed ETFs.
- For arbitrage reasons, if any, as described later.

The theory of storage applies to any commodity that can be physically stored and makes two main predictions, both related to the quantity of the commodity held in inventory (also known as stocks, a term we avoid due to its confusion with equity markets).

[...]

When there is a situation of scarcity (low inventory), spot prices will rise as purchasers bid whatever is necessary to secure supply. The effect will be less pronounced in longer term futures, since market participants know that higher price will, in the long term, stimulate increased supply and allow for a rebuilding of inventory. The effect, with *spot price > futures price*, can be extreme, and is known as 'backwardation' [...]. Conversely, when supplies are ample, spot prices can become depressed with respect to futures prices. However, this effect, with *spot price < futures price*, termed 'contango', is usually less pronounced.

(continued)

6 *Accumulation versus networking*

(continued)

[. . .] [Another important concept is] 'convenience yield', i.e., the convenience or benefit derived from holding the physical commodity rather than a paper futures contract. This was measured as a percentage yield (as proposed by Working) which the holder of the physical asset implicitly receives to offset the decline in price.

In conditions of scarcity, not only will spot prices be elevated, but they will also experience elevated volatility. This is because in a tight market, any news about short term supply, demand or inventory will have a large impact on the spot market.

[. . .] In conditions of abundance, this effect will disappear, and there will be no pronounced difference between the volatility of spot and futures prices.

Source: Geman and Smith (2013): 19–20.

From a rational point of view, storage is not necessarily negative; its economic viability depends on the costs of other factors, such as transport. If transport costs are very high, it is more convenient to acquire a large amount of commodities at once and store them carefully, rather than acquiring them by a means of transport whenever they are needed. From a market point of view, assuming that the value of things is established by the meeting of supply and demand, the capacity of suppliers to store goods in suitable conditions can alter the final prices, threatening the optimal allocation of the goods. But, as we have seen, this depends not only on the possession of goods storage techniques, but also on the financial capacities of the supplier. Consequently, that economic actor can survive in the market without immediately selling his goods because he has a large amount of current assets (liquidity). Furthermore, storage can be a sophisticated *strategy* – included in the range of individual rational actions (Baert 1998) – when it concerns the prevention of action by others. If a supplier forecasts a turbulent provision market for his means of production, he can choose to hold back part of them instead of inserting all of them in the production line. Security of provision can be a better strategy than producing and selling as much as possible at once.

Strategic behaviours in the field of goods storage can easily become speculation on the one hand, and hoarding on the other. Speculation, as mentioned above, is a sort of hyper-rationalization of economic behaviour, an excess of strategy, a desire to exploit a good without limits, thanks to a dominant position in the market. Hoarding is a hyper-irrational behaviour dictated by an unjustifiable fear of penury in the near future. Storage appears to be analytically midway between these two behaviours:

Speculation ——————— Storage ——————— Hoarding

An astute combination of keeping and leaving, maintaining and consuming is in any case economically very risky, as well being a socially reprehensible behaviour. In the same way, hoarding is risky for the finances of hoarders and is a symbol of a lack of confidence in society. It may be a symptom of mental disorder. Storage is in the

Accumulation versus networking 7

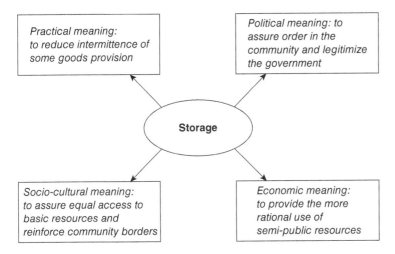

Figure 1.1 Polysemy of storage practices

middle; theoretically, it is a rational action deeply affected by the social environment. However, there is no a sharp division among speculation, storage and hoarding; the borders are contingently established in particular situations, according to social pressures and specific events. In this sense, storage is visibly a *social practice*, a complex of semi-spontaneous actions usually integrated in a script or a carrier, but sufficiently flexible to be used in changing situations (Hargreaves 2011). In conclusion, storage is not an automatic behaviour nor a pure rational choice, but rather a package of actions with an internal, adaptable theme that recalls the human archetypes of our ancestors (see Figure 1.1). It therefore has a very remote origin, embedded in changing settings and asking for impulsive decisions.

1.3 Why storage is once again important

The first section listed a number of reasons why storage has been a hidden or neglected social practice in the industrial era. They can be resumed in new terms in order to show their potential weakness: organizational reasons (development of global networks), technological reasons (development of efficient conservation techniques) and power reasons (availability of cheap energy). There would be a further cultural reason: storage in a world conceived in continuous change becomes the symbol of a static, immobile, outdated condition. This belief is a modernity framework that has affected scientific and technological research as well – a sort of epistemological prejudice against storage practices that social sciences can contribute to unveiling.

The three most immediate reasons for obscuring the importance of storage have recently changed. The first of them – globalization – seems to be an endless process:

> International trade flows have increased dramatically over the last three decades. According to WTO trade statistics, the value of world merchandise exports rose

from US$2.03 trillion in 1980 to US$18.26 trillion in 2011, which is equivalent to 7.3 per cent growth per year on average in current dollar terms.

(World Trade Organization 2013: 55)

Globalization is a trend with few static moments, because the decline of some countries has been more than offset by the rise of new ones in the world scenario:

> Central to this development – and its continuation – is the unfolding *death of distance* and the on-going transport and communications revolution that lies behind it. China could not have become the new 'workshop of the world' without the transpacific 'conveyer belt' provided by breakthroughs in containerization after the 1970s. India could not be a new global services hub without the invention of fibre optics and broadband.
>
> (World Trade Organization 2013: 55)

Despite this dramatic growth, even the experts of the World Trade Organization wonder whether:

> transport and communication costs continue their dramatic, linear decline as a result of continued incremental technological improvement or even the introduction of entirely new technologies? Or will marginal improvements begin to diminish in the future, making declining transport and communications costs a less salient shaping factor for world trade – even leading to a slowing of trade growth?
>
> (World Trade Organization 2013: 55)

Storage enters precisely at this point. Can we expect the endless development of trade in everything – the symbol is an item shipped by a drone – the development of a world network of rapid exchanges so that large and permanent deposits of goods are useless? Or, on the contrary, can we expect that increasing costs of transport, geopolitical turbulence and high environmental impact of movements will boost local economies that hold large reserves of materials? For example, the food staples market operates as a giant network able to compensate for the weak production of one region with the flourishing of another. With rapid movements of cereals or other staples like soya beans from one corner of the world to another, the urgency of stocking foodstuffs is very low.

But if this mechanism misfires – for example, because transport costs become unsustainable – every country must seriously consider its reserves of basic foodstuffs. Storage becomes a strategic option for governments engaged in the task of providing last-resort goods. Furthermore, the globalization of exchanges is not always the best solution for the distribution of goods. We already know that for some goods the market fails; thus, the globalization of water provision encounters severe obstacles because huge amounts of water are not easily transportable, and in any case they are linked to rights of local people and the needs of the environment. Even with the best technical solutions, it would be impossible to move drinkable water throughout the world. Moreover, long-distance transport creates quantitative and qualitative losses. Seasonal foodstuffs are moved from one hemisphere to the other so that consumers can have access to them all year round. But again, the environmental and economic costs of this commerce are very high.

Storage as an alternative to the just-in-time market for every kind of good again appears a reasonable way to ensure a certain continuity of provision during the year. Stored ultra-heat treated (UHT) milk is not like fresh milk, of course, but it is a small price to be paid when globalization shows its shortcomings. Such a price can have positive side-effects: the rise of new ways to prepare goods, new consumer tastes and new products to be launched on the market.

But storage can also be conceived as an adaptation to climate change (International Water Management Institute 2009), countering the impact of peak and trough events. Some studies shed light on two factors: the increased variability of climate affecting the world production of agricultural goods, and the increased demand by affluent people for richer foods like meat (McMichael et al. 2007). But we can add a third factor: the increasing use of biomass for energy purposes, which subtracts agricultural land from foodstuff uses. We may draw a parallel with water: climate change and the increased demand for a plurality of uses makes water an increasingly scarce good (along with the opposite situation of floods). All these factors lead to the same result: a temporary crisis in the provision of vital resources (Goldenberg 2012), affecting areas less able to offset decreased availabilities with either external purchases or an internal increase in production.

Storage is not *the* solution, of course. But it can be included in a *set of practices* that aim to adapt to climate change. Water is a paradigmatic example. If there is irregular rainfall – strong storms alternate with periods of drought, even in temperate areas – because of climate change, the increased capacity to store water can stabilize its provision across the year (Eguavoen and McCartney 2013). In the case of food, the situation is more complex because food has long been a commodity usually placed on the market. Variations in production are compensated by changes in the places of provision, assuming that they are free of customs barriers. But this is indeed the problem. Storage in the event or prediction of food shortage is a public measure; it becomes a rule external to the free market, usually imposed by a public authority. Storage shows here all its political nature as an enforcement scheme. It is less a physical place to store items than a regulation of the market, a subtraction of goods from the market in the event of an established or forecast crisis. This regulation means there are public mechanisms of redistribution and compensation for entrepreneurs unable to sell their products (see Section 2.2).

Storage can also be seen as an emerging practice in everyday life. Studies on consumerism have shown three bad habits in developed countries: excessive consumption, wrong composition of consumed items and too high a production of waste. Storage again is not the 'magic wand', but can be rightly included in practices to limit bad habits. In some respects, consumption can be seen as poles apart from storage, for it contains the idea of the item's immediate destruction. This is clear for fresh products, while it is less comprehensible for durable goods like pasta, for example. The tendency to buy things only to enclose them within the home is a form of compulsion similar to the uncontrolled urge to consume. In this sense, consumption and storage are pathological manifestations of the same tendency. They can be framed as a vicious circle of dependency on things (see Figure 1.2).

But storage is seen here as the capacity to moderate and plan consumption in order to have the regular availability of a good over time. It is a highly rational practice because it involves *planning*. Items are not bought according to the impulse of the moment or because advertisements have been successful, but according to a rational account of what is necessary in a certain time span. Thus, storage usually averts the first criticism levelled against Western-style consumption. In the same vein, the composition or quality of goods are carefully evaluated with a view to storing them. Storage is an act of conservation where the mix of ingredients must be attentively calculated. For example, fruits must be stored with an exact proportion of sugar or alcohol, otherwise their long-term conservation risks diminishing or destroying their intrinsic qualities. It is therefore likely to prevent the purchase of unbalanced quantities of certain items. Of course, we could introduce considerations concerning *dietary trends*, which are nothing other than shrewd calculations of food mixes (Fargeli and Wandel 1999). The third way in which storage prevents bad consumption habits concerns waste. The best technique to conserve goods entails recycling, re-use, or simply complete use at the right time. Left-over bread can be transformed into a cake, stored in the freezer or consumed later. All these are in a broad sense storage activities; they once again show that storage is a package of actions to be incorporated into long-term habits or lifestyles.

Storage, seen as a bundle of activities, has been revalued by consumerist movements for its capacity to create familiarity with practices of self-containment,

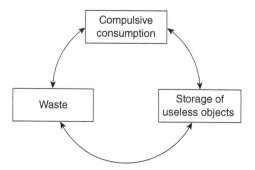

Figure 1.2a Vicious circle of consumption

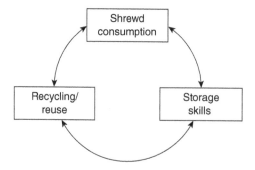

Figure 1.2b Virtuous circle of consumption

reflexive consumption and creative re-use of things previously considered as pure waste (Sassatelli and Davolio 2010). Storage, with all its manual components, is furthermore an extraordinary occasion for learning by doing – a modern pedagogy applicable to every level of education. The emerging properties of storage demonstrate the promising qualities of this practice in a world considered increasingly uncertain and ecologically endangered. In its moderate form it is an alternative to compulsive consumption, and entails a large amount of *reflexivity* (reasoning, planning, self-learning, recycling and so on). These two properties – *temperance* and internal mental dialogue – may be of crucial importance in the future. The capacity to store things properly yields important results at both the objective and subjective level: it allows a substantial reduction of total material requirements and provides a sense of security about the future. People feel that they can master activities useful for themselves and for the environment. Thus, storing provides an updated answer to the anxiety of modern people for total security: it can be the right mix of practices that helps a mild and ethically more acceptable control of the environment.

1.4 Social analysis of storage

Storage is a central topic in anthropology. The classic argument is that storage is part of a revolution that increased the productivity of agriculture and the capacity to set part of the product aside. According to Gordon Childe (1950), this happened in different parts of the world during the Neolithic period. Storage is the instrumental device that helps the passage to a new phase of *civilization*, a term later changed to 'socio-cultural revolution' (Smith 2009). In any case, it is a term conceived in two senses.

The first concerns sedentary settlement and concentration of the population. The chance to store food and water enabled people to stay in the same place almost permanently, because they had nearby the basic resources for living. But storage worked immediately according to scale economy, and it was probably convenient to store items in larger containers. This interacted with the tendency to live in proximity to other persons. The consequent concentration gave impetus to the growth of towns. The second sense concerns power and stratification. Large containers of food and water require central control and formation of the related governance apparatus, which gives rise to social stratification that is more complex than clan relationships (Woodburn 1982). The presence of a storage system also means the formation of a surplus of basic resources able to nourish new roles and functions, and this was the origin of a more sophisticated stratification within the community based on the prestige of the function and on the capacity to manage the surplus.

The causative force of surplus is arranged in clear sequence by Bruce Owen, who notes:

- Some or all of this surplus is collected from the farmers, stored, kept track of, and redistributed
    - some of it may be a safety net for farmers in bad years

- some of it (in reality, most of it) is distributed to specialists who do things other than produce food[:]
    - craft producers
    - priests
    - administrators [. . .]
- this happens in central locations
    - most notably in centralized storage places like government warehouses
    - under the guidance of community leaders
- who become more powerful due to their control over the stored surplus.

(Owen 2008: 1)

This reasoning has been applied to water management as well. According to Karl Wittfogel's (1957) notion of *hydraulic civilization*, small farmers submit to a leader in order to build and maintain large-scale irrigation works which increase their production and protect them from flooding. These irrigation systems demand a high level of coordination that only a central authority is able to provide; in fact, it has to engineer the projects, form and coordinate working groups, and supply the workers with materials and food. All these activities require a strong and bonding leadership which gives rise to the same centralized government apparatus seen for the theory of surplus.

The hydraulic hypothesis highlights a specific aspect. An irrigation system with channels, dikes and artificial ponds is a storage system organized in the form of a network. Water can be conserved not only in containers, but also in more distributed and connected lines. Not only food, but also water was a source of power whose management required direction, expertise, coordination of work, and internal and external control. All these functions needed specialized personnel or dedicated officials. A similar line of reasoning has been adopted for energy, and especially for fire. When it was discovered how fire could be managed, there came a turning point in civilization with the rise of a group of energy managers, the first nucleus of a stratification of power (Goudsblom 1995).

Indeed, civilization approach has been criticized by the same anthropologists (Scarborough and Vernon 2003; Mollinga 2008; Agar 2012). The previous explanations contain the implicit assumption that there is a linear progression of levels or arrangements from simpler and poorer to more complex and richer – a consideration immediately applicable to storage:

> Modern storage and transport may reduce vulnerability to natural crises, but they increase vulnerability to disruption of the technological or political and economic basis of the storage and transport systems themselves. Transport and storage systems are difficult and expensive to maintain. Governments that have the power to move large amounts of food long distances to offset famine and the power to stimulate investment in protective systems of storage and transport also have and can exercise the power to withhold aid and divert investment.
>
> (Cohen 1989: 134)

Thus, storage, like other archetypal organizations, cannot be conceived as either a progressive system or as a certain turning point in society's evolution. Its co-evolutionary role is disputable because it embodies a rigid and unilateral combination between social forms and technical devices. The evolution of storage techniques is not a sure sign of progress or civilization. This introduces a more subtle perspective on the relationship between the technical and symbolic dimensions. If we uncouple the link between technical (storage) and social (affirmation of a central government), the boundaries of the two fields are also put in question (Allenby 2011: 11). In other words, storage is no longer seen only as a linear instrument of improvement, but assumes eccentric meanings in relation to social and cultural aspects.

Julia Hendon (2000: 44–45) argues that storage, as a component of mutual knowledge, becomes 'part of the complex relationship between human actors within a landscape that they create and inhabit' and 'acquires a moral dimension because it is part of the process connecting resources with people's needs and desires'. She shows how storage pertains to the sexual division of labour, and how its often hidden nature – for example, underground container management – is occasion for distinguishing the knowledge levels in the community. In sum, the typical causal sequence 'improvement of agriculture → surplus by storage → social differentiation' is blurred: some practices of storage are not pure instruments of a more functional society, but they have been introduced to establish cognitive or social hierarchies. If so, *the ritual functions of storage may sometimes precede its instrumental function*. The most prudent interpretation is to minimize rigid linear explanations and combinations, and view storage as a complex activity in which instrumental and symbolic meanings are intertwined in the history of a population. That, indeed, is the result at which sociology arrived for many activities with high content of science and technology (see Pellizzoni 2015).

This last statement introduces two further social approaches to storage. One is linked to consumption styles, the other to socio-technical systems. The former pertains to the vast area of consumerism. It is surprising how little attention consumption studies have paid to the theme of storage. Nevertheless, some insights are provided by studies based on the relational nature of consumer identity (Cherry 2006; Sassatelli 2015). In that perspective, consumption escapes the double trap of being seen as a pure instrumental act guided by the maximization of preferences or as a product of hidden persuasion commanded by the dominant production class (Osti 2006). For example, an identity perspective applied to the wardrobe, as a system to store clothes, produces the following analysis:

> [Although the] wardrobe is more readily understood as a definite object, I wish to argue that it also commands a set of distinctive and identifiable spatial practices: forms of structuring, delimiting, and organizing clothes, as well as the social meanings and identities articulated by these forms.
> (Cwerner 2001: 80)

Cwerner uses Goffman's concept of the *identity kits* that 'consist in various objects such as clothes and make-up equipment, but also an accessible, secure place to store these supplies and tools' (Goffman 1965: 246) to point out how consumption reflects the will to self-distinguish – called identity – and to communicate with others in order to feel similar – called identification (Gallino 1982).

The analysis of the wardrobe as a sort of reserve for identity raises another aspect of consumption even more neglected than storage: the idea that some items are purchased solely to remove them from the view of the others; they belong to a sphere of intimacy, privacy and familiarity which forms that part of the identity to be hidden from others, or better, to be shown on special occasions. Stored items have this character. Storage, when it is not the pathological collection of unused things (compulsive hoarding, or *disposophobia*), allows this game of hiding and showing objects that reflects the previously mentioned dialectic between identity and identification. In such a way their value increases, as well as that of the special moment when they are shown or consumed. Such behaviour recalls the idea of the intrinsic pleasure of knowing that one has something recondite. Of course, it works better with immaterial goods like memories and feelings; but the idea of a remote place where one's most valuable things are conserved has fascinated generations, as the possession of never-shown jewellery demonstrates.

Despite these important insights, the identity perspective risks overlooking other important aspects of consumption like skills and relations. Both matter a great deal for storage. Skill is competence in all the phases of consumption, including the possible moment of storing items. This ability is crucial for the immediate aim of using an item correctly, but it is also important for the intrinsic satisfaction it provides. Making tomato sauce is useful for a good meal in winter, and it gives the pleasure of being able to carry out the work. This is the essential motivation not only of food storage, but also of a wide range of do-it-yourself activities (horticulture, model building, self-built furniture). Storing is therefore part of the broad field of practices in which different types of knowledge are involved. Ability and prestige are mixed in the capacity of stored items to be shown and used at the right moment.

Relations also matter in terms of storage, for the reasons just mentioned. Not only can storage be a collective endeavour, but the stored items can be consumed together. The conservation of traditional foods like tomatoes and salami has long been a collective ritual open to people from outside the family. In more theoretical terms, relational goods are produced together with the storing of material goods. They open a breach in the dominant and individualistic perspective of rational choice (Uhlaner 1989). But the challenge is to determine how much of the traditional way of storing has remained, or whether it has assumed new forms in what can be called *retro-innovation* (Stuiver 2006). The relations are not always good and produce conviviality; in some cases they are conflictual and require a great deal of mediation; in other cases they have been codified in special institutions, as the history of water management shows. If we adopt an *open* view of relation – as Godbout and Caillé (1998) say of modern gift-giving – storage is seen as a set of rules constantly adjustable according to people's contingent interactions.

However, a relational approach to delayed consumption – another way to label storage – risks being too situational, too historical, too linked to special cases.

A correct practice-based approach searches for a minimal *set of regularities*. This does not mean finishing with rigid rules of behaviourism, whose outcome is the maximizing actor. Differently, the practices approach looks for sets or packages, variably combined, used in certain circumstances. Storage can be stylized from this perspective as follows: in response to the possibility of a natural disaster, people adopt different packages of measures according to a plurality of factors – personal culture, social *embeddedness*, degree of exposure to the media, location of their physical goods, prevention devices offered by the community and so on. Thus, storage becomes *preparedness*: entry to an institutional frame where a set and a sequence of actions are forecast and organized.

Numerous examples will be provided in the subsequent chapters. What is to be highlighted now is the institutionalization of storage practices. These are codified in state rules, specified in regulations, organized and explained by agents, learned and implemented by ordinary people, and verified by evaluators. Institutionalization is therefore a long and intricate process, whose results may be very different from those initially thought or enforced. Cognitive institutionalism insists on the need for *legitimization* of the bodies belonging to the same organizational field (DiMaggio and Powell 1991).

If we consider the technical dimension of storage more closely, the literature envisages the codification of stages or levels of diffusion, the *transition theory* of Geels and Schot (2007), for example. Storage is an activity subject to constant innovations; however, specific groups of actors – for example, public agencies, universities and private companies, the so-called *triple helix* – are able to develop a dominant research trajectory, unable to lock out the storage technologies (Lehmann et al. 2012). The social sciences adopt a critical attitude toward expert circuits, seen as power groups able to use material and immaterial resources with the intent to impose their research path (Viale and Pozzali 2010). Again, this is a matter not only of the number of innovation adoptions – generally seen as the only parameter of technical scientific activity – but also of contingent fortunes or astute communication strategies. This brings us the concept of the *technological fix*, which is a package of procedures recognized as the only way to deal with a problem (McEvoy 2014). In our case, we can think of a *storage fix*: a standard set of rules for storing goods that dominates the panorama of production and application.

Of course, we can lament the opposite of a storage fix: a confusing and uncertain world of applied research consisting of many unreliable companies and developers unable to indicate a solution to an urgent problem. Reactions to disasters or epidemics often reflect this pattern: each country or authority goes its own way without a clear definition of the issue. Carbon capture and storage – a topic not developed in this book – is probably in the same situation: *a clumsy solution for a messy problem* (Verweij and Thompson 2006; Markusson, Shackley and Benjamin 2012).

This introduces the last body of literature useful for understanding storage: *risk management*, which can be considered as social practices (Sarkis 2006). It consists of packages of procedures adopted to deal with highly uncertain events

(Kasperson et al. 1988). First, there is the phase of analysis of the situation, of a historical series of events and of resources. Then there is the organizational phase, when sequences of actions – monitoring, preparedness, emergency reaction and so on – are envisaged and inserted in protocols; finally, there is the phase of formation and communication to the people potentially affected by the negative event. Storage is mainly a sub-phase of preparedness (storage of food, medicines, fuel and so on). This practice must be analysed and communicated as well; hence storage is part of the entire process of risk management. We could add that storage itself can be a dangerous activity that requires the application of risk management procedures. The large fuel storage tanks used for the prevention of energy shortages are usually subject to careful protection measures.

Risk analysis in social science has evolved from simple psychological support for the application of mathematical forecasting models, to a cultural perspective in which cognitive frames are involved in the negotiation of risk definitions among experts. The bias is strongly constructivist. Accordingly, storage as a precautionary activity is a factor of central importance. The functionality of a depot, the amounts to be stored and the ways to keep the storage structures efficient are all elements that enter experts' calculations on damage prevention or mitigation. Energy, especially electricity, is very difficult to store; but it is crucial to evaluate whether and how to store it in case of blackouts.

Energy storage will be the subject of Chapter 4. Nonetheless, it is important to outline tools offered by cognitive sociology, such as the capacity to frame the storage issue within territorial levels of analysis, technological trajectories, prevention or anticipation policies, and cultural attitudes of this or that population. Storage is thus not only a measure of civilization, but also (part of) a cultural landscape and a practice included in social networks. Local expertise in storage has been swept away by excessive standardization of construction methods and ways of dwelling. This is evident in the destruction of every storage device traditionally found in Mediterranean houses (Conte and Monno 2001). Nevertheless, the issue is not a case of simple revenge of traditional ecological knowledge. Nowadays, storage is not imaginable without new appliances and procedures.

Food, energy and water risks are increasingly difficult to evaluate at local level because of their mobility; a mix of cultural frameworks and levels of analysis is therefore necessary. Storage, with its strong physical dimension, has the feature of being understandable and manageable at different territorial scales. Studying things with the intent of storing them can therefore provide a good opportunity to form a new alliance among *knowledges*, combining the most sophisticated and abstract with the most simple and locally rooted (Hernández-Morcillo et al. 2014).

In order to place storage in a full social perspective, the final step is to conjugate different cultural frames – expertise included – with a primordial contraposition between accumulation and networking or storage and reciprocity. We will make this attempt with an old anthropological research study on an unlettered population in Africa. The point of departure was how the people reduced risk. One way was to develop a wide network of relationships based on reciprocity:

Accumulation versus networking 17

In Botswana, self insurance is practised by cattle owners who keep their herds at several widely separated cattleposts, where the chance of loss at each cattlepost is independent of the others. A reciprocity network can act in this way as well. The simplest form of reciprocity is a norm of sharing and gifting, and it acts like insurance in the way that it reduces variance in income. Because the risk is shared among a number of different individuals, each is protected from the possibility of a catastrophic loss. The 'payment' for this protection is the obligation to help when someone else is in need. It follows from the argument presented above that for generalised reciprocity to act as a form of risk reduction, the risk facing the different individuals must be independent.
(Cashdan 1985: 456)

The independence of many actors was therefore the first condition for reciprocity to be useful in facing adverse events due to drought, devastating conflicts with other tribes, or livestock epidemics. Reciprocity was seen as the opposite to storage. This is the same idea, recalled in this book, that a network is an alternative way to store vital resources. Instead of keeping them enclosed in one place, they are maintained in circle among many actors. Risk reduction is evident: a fire can destroy a cereal store, ruining a big farm for ever; if the same cereals are conserved in many distant places by many small farms, and if they agree to help each other (reciprocate) in the case of fire, the risk of famine is drastically reduced.

As sources of variability in the opposition between reciprocity and storing, Cashdan adds two geographical dimensions: *mobility* and *distance*. Mobile people, for example nomads, find it easy to establish exchanges, and as a result it is disadvantageous for them to hold resources blocked in a storehouse. Likewise, great geographical distance between points of food provision increases the costs of transport, so it is more convenient to store than to exchange. Furthermore, if the environmental risk is large in scale – for example, an entire region periodically hit by drought – exchanging within the area is useless, because all are presumably affected by the same adverse event; this is another reason to choose storing rather than reciprocating: 'We should therefore expect storage to replace reciprocity as a means of risk reduction when a population becomes more sedentary, especially where the source of risk is regional rather than local' (Cashdan 1985: 458).

Authoritative confirmation is provided by the *International Encyclopedia of Economic Sociology*, in which two entries mention the point: 'Reciprocity is the dominant form of exchange in foraging societies, particularly in those with great variation in food sources and limited options for storage, where reciprocity functions as "social storage"' (Moritz 2013: 677) – in other words, 'a diminution of food storage gave rise to a growing solidarity instead' (Mosselmans 2013: 590).

The contraposition between network and accumulation is thus enriched by several variables which provide a final framework, hopefully well balanced in terms of generality and concreteness (see Figure 1.3).

This framework needs supplementary explanations: in general, the storage contraposition is interpreted as an agent/society dualism; storage is in some way antisocial because it creates disparities and individualism among people. Some

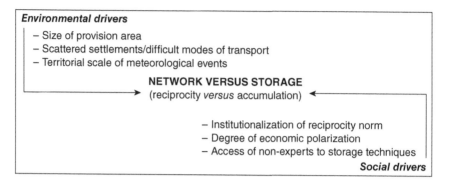

Figure 1.3 The network/storage contraposition according to environmental and social drivers

are better able to look after their resources, and this creates different capacities for accumulation, prestige rankings within the community, and chances for exercising power over the other members. By contrast, the interdependency of vital resources creates the conditions for stronger ties and a common understanding. Numerous variables enter this general picture; in the framework being considered, they are organized into two categories: environmental and social drivers. These may greatly change the picture, especially in highly mobile and ecologically stressed societies. The framework should thus be updated to contemporary situations where climate change (at the moment very evident in certain locations) and wide differentiation of storage capacity (in terms of knowledge, places, access and so on) are dramatically apparent.

2 The modularization of food processing and consumption

In this chapter, the storage practices concern food. Its preservation involves a wide range of activities, from production to consumption, from simple manual household tasks to highly industrialized procedures. All these practices are framed in four dimensions of storage: (1) security in provision, (2) safeguarding the health of people and the environment, (3) rational prevention of waste, and (4) an activity charged with a variety of symbolic meanings. The chapter concludes with consideration of the chances that food storage can provide new insights on the environmental crisis.

2.1 Food conservation, a wide-ranging and ambivalent issue

Food storage has a very long history (Diamond 1997; Higman 2011). As mentioned in Chapter 1, the storing of food is linked to the first steps of human adventure. Usually, the capacity to store is divided into three stages or turning points:

- from the era of hunters and gatherers to the era of agriculture;
- from a clan/community level of organization to a centralized authority with redistributive tasks;
- to a more rational organization of life through the reduction of dependence on natural events.

In each stage there has been a combination of technical devices with a division of labour, the assembly of tools and the assignment of tasks. Food storage is thus a fundamental component of a socio-technical package: the combination of practical methods with the capacity to coordinate people's work. For example, having olive oil available throughout the year requires the ability to make the product from olives (milling) and to use jars (preservation), together with the distribution of those tasks among the people in the supply chain. Apart from the ability to manipulate tools and people, control over olive groves and jars leads to differentiation within the community. Because human abilities and means ownership imply differential access to vital practices like storage, they entail the exercise of power. Basic factors of social differentiation like knowledge, property and authority heavily condition every storage practice. Thus, in pre-industrial society, food storage capacity was a primary source of social rank.

The above-mentioned historical stages represented by food storage can be interpreted in light of two main social processes: *control over natural resources* and *control over people*. The first point is clear: natural resources are heavily manipulated in order to achieve specific human aims; food storage makes it possible to live relatively well in fallow seasons. Thus, storage requires and allows the rigorous distribution of tasks over time. It creates regularity in schedules – a crucial condition for every organization. But storage also allows precise distribution in space. A certain type of storage, for example in barrels, facilitates the transport of food. Hence, workers can be adequately fed in those workplaces where the industrial conditions are more favourable even when foodstuffs are scarce. The modular storage of food and its transport are bound up with each other, and they allow great advances in the exploitation of natural resources. Consider mines, the timber industry or the more recent offshore oil platforms. All of these need a capacity to store and transport food in remote areas. This not only reduces the dependency of human activity on local conditions; it also greatly increases the capacity to exploit the environmental resources of those areas systematically – an old process that has recently accelerated and is the origin of the ecological crisis.

The second point – people control – is more ambivalent in its meaning. Food storage has been crucial for warfare. The capacity to invade others, as well the capacity to withstand enemy siege, are closely dependent on the victuals opponents can provide and store. The success of certain military campaigns has been linked to the amount of victuals and the ability to preserve them for a long time in adverse circumstances.

The historiography of wars rarely considers this crucial capacity to store food for soldiers, given that looting of civilians is not always efficient or possible. History concentrates more on military strategy, weapons and troop morale. Food logistics are rarely analysed as crucial for the outcome of military conflict. In the history of food canning, it is curious to note that this kind of storage was attempted by Napoleonic troops in order to ensure their capacity to conduct long and distant war campaigns. Canned food is therefore a symbol of modern wars; it represents a military technology later developed for civil purposes.

But people control by food storage is ambivalent. When food storage requires large containers and demanding construction procedures – imagine a giant granary for thousands of persons – it is evident that (a) only organizations with large endowments of resources can build and manage such infrastructure, and (b) access to the storage infrastructure must be strictly regulated. An authority, permanent surveillance and standard criteria for use are necessary. Furthermore, if food preservation requires constant and sophisticated control over possible causes of degradation, a class of experts arises (Mishra, Prabuthas and Mishra 2012). They achieve their power through jealous protection of their exclusive and valuable knowledge, which creates problems of reciprocal understanding with consumers (Krystallis et al. 2007).

By contrast, if the storage capacity does not require large initial investments, nor centralized police patrols and stringent formal access procedures, control over people is less pervasive. When food storing techniques are simple, require small amounts of energy and are easy to learn, everyone can store food at home; of course, if there is strong central control over just one of the factors mentioned – food production, energy provision, knowledge access – the entire storage process is socially unbalanced. It is thus necessary to consider the forms of control over food storage. These have been addressed within the frame of *food security*:

> Beginning with Foucault's writing on food provisioning in the mercantile period, this paper explores how a moral economy of hunger is gradually replaced by a political economy of food security that promotes market mechanisms as a better protection against scarcity. In Western Europe the emergence of political liberalism and *laissez-faire* economics substantially shaped how hunger and scarcity were conceptualised and socially managed. Beyond Europe these social forces were manifest in the development of colonial plantations. Here the transformation of non-capitalist social formations into market economies – what Harvey [. . .] terms 'accumulation by dispossession'– was a foundational moment in the development of a global provisioning system that undermined the anti-scarcity strategies of some populations, while ensuring food security for others.
>
> (Nally 2011)

This quotation inserts food storage within the current of thought based on *governmentality*, and more recently on the critique of neoliberal policies (Pellizzoni 2011a). The most interesting point is the special light cast on self-storage practices. It was stated above that control over people decreases when two conditions are in place: the modularity of storage techniques, and easy and wide access to the content and means of food storage. To return to the olive oil example, if pressing olives and making olive oil barrels are easy at farm level, given that the raw materials are available, storage evades political or central control. Of course, controls after these commodities have been produced are feasible, but they are very expensive and unpopular.

According to a governmentality approach, control is exercised in any case; but it changes its form, becoming more subtle, almost invisible. For example, it can range from the traditional cupboard lock to food safety norms to be followed while processing, storing and consuming food (Punch et al. 2009). But the control can become even more imperceptible when it is based on an assumed agent autonomy: which means, in terms of food storage, a household's self-sufficiency for an entire year. Thus, invitations to store food in order to be more autonomous in reality become fine-tuned instruments of control that are more efficient because they are not perceived as such by the population. On the contrary, people like storage practices because they believe that they are freer from external pressures. This is the heuristic of governmentality. There is an *enticement* for people to exercise self-control. The rationale of storage does not stem from considerations of self-interest, but is induced by external forces.

Defining what these external forces are is probably a weakness in the governmentality approach. We know they can be a government – that is, the officially recognized uppermost authority of a state – but they can be also the administrative apparatuses of that state, or even the minuscule processes and instruments of government (MacKinnon 2000). Even if we start with identification of a substantial or underlying *logic of the system* (Sending and Neumann 2006), we cannot arrive at a clear recognition of the *prime mover*. On the contrary, according to an actor-oriented perspective, precise identification of subjects, objects and processes is necessary. Nevertheless, food consumerism – an approach more biased towards agency than governmentality – also has limitations:

> Food consumption is a complex social phenomenon. Therefore it is not advisable for the discussions about ordinary food consumer's position in relation to sustainable consumption to understand consumption optimistically as only increasingly strong agency of the so called political consumer. Neither is it fruitful to understand consumption pessimistically as only increasingly disciplined or disoriented agency as in the governmentality argument. Both types of assumptions rest crucially upon assumptions about what other actors in society do [. . .]. And both types of assumptions are too crude in themselves to understand the varieties of processes of everyday life dealings with environmentalised consumption.
>
> (Halkier 2009)

Yet governmentality, with its emphasis on basic organization and devices of power, can be useful for understanding food storage. In fact, its line of reasoning leads straightforwardly to security, a target essentially linked to food storage capacity. Food security is a complex of things:

> Food availability is only one aspect of food security. Lee and Greif review four core dimensions of food insecurity: *consumption* level pertains to the number of meals eaten per day, the amount being eaten, and the degree of regularity of meals; *quality* refers to both the nutritional aspects of food and personal, subjective preferences; *sources* indicates both the foundations from which foods are supplied and the personal and cultural acceptance of the sources; and *cost dimension* is central to fully considering components that compose food security/insecurity.
>
> (Haering and Syed 2009)

According to this broad definition, there are three main criteria which identify food security:

- availability – healthy food with sufficient nutritional capacity;
- stability – permanent provision despite seasons, disasters and crises;
- accessibility – without discrimination of any kind, not least against poor people.

But the concept of food security can be enlarged further to include considerations of the food system required to provide security: the absence of oligopoly among farms and food processing firms, democracy in the decision-making chain governing the food system, and environmental sustainability (Carolan 2013). These considerations raise a further question: what is the territorial dimension of food security? Usually, this is a matter for national authorities; but it would be interesting to verify whether such competencies are decentralized to local authorities (Hinrichs 2013), or whether it is necessary to go beyond the national dimension because it too narrow in the face of global environmental threats linked to food provision (see the emblematic case of peak phosphorus in Cordell, Drangert and White 2009).

To return to the general issue of storage, control over people can therefore come about through the imperatives of food security. In order to ensure available, stable and accessible food, some of it must be stored. Thus food can reach remote human settlements (availability), it can be consumed when immediate sources are not ready (stability), and it can be easily distributed in its modular forms to poor people or to people affected by disasters (accessibility). How food is stored is a crucial measure for guaranteeing social security.

Food security can therefore be inscribed according to the logic of governmentality: its purpose is to ensure the population's well-being, it prevents disorder, and it is achieved through apparatuses and procedures. One of the latter is storage, which must be safe, efficient and free from discrimination of any sort.

Nonetheless, considering food storage as a matter of control risks neglecting the elements of agency, freedom in the symbolization process and the capacity to resist any top-down pressure (Maxwell 1996). Thus, food storage can be seen, on the one hand, as a *symbolic activity* whose aim extends beyond the protection of food, and on the other, as a way to create a *more rational use of food* in a complex organization of life. They are both long and robust traditions of thought, often seen in contraposition by social scientists (Kaufman 2006). An application to storage has not been developed, but on food the literature is definitively broad and inclusive (Sobal and Bisogni 2009). Meat consumption is a typical case where the symbolic meanings of strength and prestige are confronted with rational evaluation of its nutritional performance, healthiness and environmental impact (Beardsworth and Keil 2002: 193ff.).

As regards symbolization, food storage can be part of a lifestyle motivated by religious imperatives, a need for distinction in a mass society, deference to an authority, or a signal of well-being. Many examples can be cited for these motivations: the habit of storing food and other basic goods among *Mormons*; the stored food linked to different cooking styles and *gourmet groups*, with their rituals, guidebooks, famous chefs and prestigious national cuisines (and now television cookery competitions), and the recovery of local or traditional food shaped by storage recipes, whose adoption is a clear signal of loyalty to a movement like *Slow Food*. Finally, because food preparation and storage are place-based activities, they positively affect the attachment to place, providing a sense of well-being (Macintyre, Ellaway and Cummins 2002).

Symbolization is a social process that applies to every phase from food preparation to consumption. Procedures of control are framed by cultural values. Healthiness, especially hygiene norms, can be considered to be largely affected by *local* principles. Thus, the powerful mechanisms of food authorization and control – not to mention certification – can be viewed as a gigantic ritual in which single, more scientifically evidenced procedures are inserted. Storage practices are therefore a mix of tested procedures and moral beliefs. They themselves are objects of symbolization, of meaningful *boundary work* (Gieryn 1983).

As regards more rational use of food, storage has an important role in attempts to reduce waste along the chain of food production and consumption. If we assume that producers and consumers are rational actors, they will seek to reduce costs as much as possible, in particular those due to malfunctioning systems. Industry uses storage as a means to optimize transport, to have the appropriate quantity of items available to be processed according to the machinery's needs and labour availability (for example, lean production). In other words, food storage represents perfectly scale economies. Storage capacity is also crucial for selling processed food: stored food addresses demand peaks, the objective of maintaining market quotas, and the supply of or demand for highly differentiated kinds of food. Food storage then enters harmoniously into marketing science and practice (Clemons and Santamaria 2002).

The rational consumer seeks to avoid wasting food, by buying long-preservation items like canned or frozen food, or food that can be easily recycled or preserved after partial use. Wasting food is considered to be either irrational or immoral; there are thus many reasons for avoiding it. Unfortunately, we know this is not the case in practice, because a great quantity of food is wasted every day, especially in the ultramodern system of distribution. Hence, the tension between the rationality of food consumption and the gaps in the food distribution system has to do with the storage capacity of producers, retailers and final consumers. A *chain* seems to be the appropriate representation: a long sequence of interdependent actions, a flow of food, which should move efficiently from one step to the next, but in fact in every phase is subject to side-effects, organizational fallacies, misunderstandings and random conjunctions whose final result is a great amount of food waste. What is at stake is the degree of rationality of each actor in the chain, and the capacity of the food system to coordinate these phases or steps.

Sources of criticism are evident: the production side accuses consumers of being irrational, of buying the wrong items and quantities, while consumer associations accuse the food industry and the distribution system of providing those items that ensure the best profit, and cynically considering a degree of food waste to be normal. Thus, new intermediate organizations arise to attend to the last links in the food chain. They try to reduce waste in three ways: (1) a more flexible policy regarding food consumption deadlines, (2) better communication among stakeholders, and (c) the channelling of close-to-expiry food to charities.

At this point we can summarize food storage practices according to two metavariables: the control/agency duality, and purposive versus value rationality. The former is inspired by the above-mentioned debate between Foucault scholars and

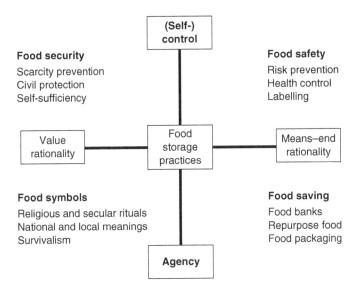

Figure 2.1 Classification of the main forms of food storage according to the intersection of (self-) control versus agency and value rationality versus means-end rationality

those who tend to favour an actor-oriented approach (Halkier 2009); the latter is inspired by Weber's well-known typology of social actions, which makes it possible to distinguish between value rationality and means–end rationality (Weber 1922). Hopefully, the first and the second couple of meta-variables should discriminate among the great variety of forms of food storage. The further operation is to cross-reference the two meta-variables to arrive at four types of food storage practices: food security, food safety, food rituals and food saving (see Figure 2.1).

2.2 Food security practices

Food, like energy, underwent a revolution when transport technologies and the openness of markets allowed stored items to be moved easily. The capacity to store food increased and changed its form: from many small containers in a short production-to-consumption chain, to a much longer chain in which the intermediaries greatly increased their power due to better transport, storage and processing conditions. The globalization of food has been the result of liberal or neoliberal policies favouring trade; but it is also the result of the increased capacity to store agriculture staples on means of transport. Granary ships are one concrete symbol of this revolution of *storing by moving*.

Large ships full of cereals arrived in western Europe from granary countries as early as the nineteenth century, provoking great changes in social and productive structures. Italy is a case in point; after the country's unification in the second half of the nineteenth century, the arrival of wheat from Canada and the USA caused a severe crisis in local production; prices decreased greatly, impoverishing farmers and farm workers – especially in southern Italy – who specialized in durum wheat. Ships worked as giant storage devices, allowing traders to wait for the right moment to sell and deliver the wheat.

A similar process happened with wheat flour: when well-industrialized countries were able to provide and transport/store great quantities of flour, the social and economic situations of recipient local economies deteriorated because their grain mill industries declined. Introduction of steam-powered roller mills in the mid-nineteenth century led to two noteworthy results: the possibility of milling greater quantities of wheat (concentration of mills), and improved capacity to transport flour because it rancid fats it contained were removed by a new process that separated elements of the grain (Perren 1990). Curiously, these industry-side advantages did not represent an improvement in the nutritional qualities of flour, and then of bread. The same happened some decades later with hydrogenated fats: these allowed the longer preservation of food, then its easier industrialization and transport, but they endangered people's health.

In sum, storing by moving had several consequences:

- longer supply chains;
- concentration of industrial processes (a smaller number of bigger companies);
- crises among local closed chains and their traditional storage systems;
- ambivalent effects on food quality.

According to scholars of neoliberalism, these trends can be summarized as the development of *processed food regimes* (McMichael 2009). Food storage has become an essential practice in the food industry, which can transform food because it is able to store it, or better put, is able to modularize its storing and processing. It is possible to industrialize Christmas cake because flour, butter and sugar – three already transformed ingredients – can easily be stocked in great quantities months before Christmas. Then planned production can provide processed cakes which will last some months. By contrast, fresh patisserie can last for only two or three days; its production is notoriously concentrated into a few hours before consumption. It can be industrialized in only some minor phases, and has to be assembled in small nearby workshops.

Again in terms of the critique of neoliberalism, 'faith in the market to deliver cheap calories continually to the world's hungry is so great that countries have been instructed to abandon longstanding practices of surplus storage' (Carolan 2013: 26). What should be added to this quotation is that storage has not disappeared; rather, it has changed form, from local and widespread to supra-local – often supra-national – and concentrated in large companies able to manage the storage by transport system.

Cereals and legumes like soya are now also global foods because they can be adapted to the virtuous combination of transport and storage. Hence, their transformation can be localized wherever it is most convenient for the food industry, regardless of how close that is to large consumption markets.

The storage/transport combination highlights another crucial aspect of the global food system: the specialization of poor countries in particular crops. According to the food regime perspective, agrifood multinationals should develop only conveniently stored-by-transport products in countries with low internal demand. This responds to the rule of the maximum exploitation of poor countries' primary resources, whose transformation happens elsewhere.[1]

It is interesting to assess the fate of different tropical or subtropical fruits according to the difficulty or simplicity of storing and transporting them as fresh produce. In the event of major problems, industrialization of the produce probably started close to the place of cultivation, reducing the crude exploitation of raw materials in the poor countries. Thus, a lesser capacity to store fresh agricultural products induced the first in situ industrialization of poor countries.[2] Nevertheless, the more fine-tuned transformation takes place in the richest countries, once the tropical product has been crudely processed in the place of origin. This seems to be the fate of certain universally popular products like coffee and cacao. First they are processed and stabilized in their places of cultivation, while respectively the roasting of coffee beans and the production of chocolate are undertaken in the rich and well-equipped countries of the global North. In the evolution of these chains, storage is not eliminated, but changes its form. It becomes a typical *modular process*, given that each independent unit (or *module*) concerns either the integrity of a good/service or the block of inter-firm relationships established to create it (Gereffi, Humphrey and Sturgeon 2003).

The progress to modularity in terms of storage location and of market range can be represented as follows:

Storage location:
Household/farm level → *Centralized large stores* → *Modular units*

Market range:
Closed local market → *Regional/national* → *International*

There is a parallel between the forms of food storage and the breadth of markets: in local subsistence markets, storage initially took place mainly at the level of the single production unit, the farm or the family clan; as the market enlarged, large

---

1 The storage phase is an aspect that is neglected by food regime analysis. According to Burch and Lawrence (2009: 268), the contemporary third food regime is based on multinational companies' retailing of fresh and health food and on 'financialization' of the entire agriculture sector; 'freshness' is not mentioned as the opposite feature of stored food, but only as a marketing label.

2 The hypothesis is striking, and it inspires a classical line of action by fair trade organizations. However, the evidence seems to indicate a more balanced situation: for coffee, 'around 40 percent of the final product price accrues in developing countries; [. . .] it is notable that a similar ratio exists in deciduous canned fruit [. . .] and in fresh fruit and vegetables' (Fitter and Kaplinsky 2001: 71). Fitter and Kaplinsky maintain that differentiation of the coffee supply undoubtedly exists, but it benefits only consumer countries, which are the final transformers, and the retail sector.

stores under the control of a central administration emerged. With the combined development of international markets and industrialization processes, storage became multi-level and multidimensional: a plurality of storage forms at different points along the production chain. This also happened because storage technology became modular: the freezing process was developed at many levels, in fixed and mobile places, from the mini-bar in a luxury car to the transoceanic factory ships able to process and freeze fish catches.

The intertwining of storage devices and food processing yields a more balanced view of the above-mentioned North/South divide in food industrialization. Storage technologies are ambivalent: on the one hand, they increase the ability to transform raw agriculture products in situ; on the other, they allow the complete delocalization of all food processing, thereby increasing the uneven levels of development. Storage techniques are therefore instruments for different paths of development. But it is reductive to view storage as a purely interchangeable instrument in the hands of ruthless speculators. It is in any case a device that is able to develop its own rules: in other words, to become *self-referential*, as exemplified by the *miniaturization* of some food storage devices (Maestrelli and Della Campa 2011). Consider the range of freezers and refrigerators involved in all the phases of food processing, from collection in the field to final consumption. When a food processing company has organized its activity according to the refrigeration chain, all changes in techniques and places of production will be affected (Pacyga 2008: 155).

The general lesson is therefore clear: food industrialization and the combined techniques of food transport and storage have created the basis for an extraordinary globalization of food security. Food provision remains constant almost everywhere despite production shortages in single localities. Modular systems of food production give great flexibility to the market – a global structure able to serve every corner of the world with the right sizes of packaged food. In fact, to use Amartya Sen's (1991) famous dichotomy, security becomes a problem of entitlement, not of provision. Processed, packaged and stored food arrives everywhere (good provision), but people have different capacities to acquire such food, for a variety of reasons to do with their social circumstances (poor entitlement). Food security is a problem of poor people or weak economies at the mercy of industrial-commercial oligopolies (Shiva and Bedi 2002).

It is so true that provision is no longer a problem that developed countries have almost completely abandoned the traditional food storage practices used to ensure essential food supplies during periods of shortage. High levels of production made possible by the selection of seeds and chemical fertilizers help to generate full food provision: in the developed countries, food is abundant, rich in calories, varied in form and taste. In those countries there were no food shortages after the Second World War. Thus, maintaining an expensive public storage system seemed pointless.

The history of the European Union (EU) is emblematic of the decreasing importance of food storage. The publication for the 50th anniversary of the Common Agriculture Policy set out an eloquent chronology in which 1962 is described thus:

The common agricultural policy (CAP) is born! The essence of the policy is good prices for farmers. With every passing year, farmers produce more food. The shops are full of food at affordable prices. The first objective – food security – has been met.

(European Union 2012: 3; bold in original)

The social climate in which the Common Agriculture Policy was born is clear: no more food shortages thanks to an efficient agriculture sector which would ensure fair incomes for farmers. The first aim (food security) was eclipsed by the extraordinary productivity increase of European agriculture, which within a few years suffered from overproduction. The other two aims – modernization of the primary sector and increased income for farmers – remained pre-eminent to the present day, even if they were affected by further aims like market inclusion and (limited) attention to the environment.

The main instrument of EU food security has been storage. This is nothing new. In Italy, even before the Second World War the storage of cereals was made compulsory in order to limit speculation on prices during periods of food shortage. It was used in the case of excessively high prices to calm the market. Today, there is still an EU public storage policy, but it is residual and has the sole purpose of protecting the revenues of farmers. In fact, it is classed as an *intervention* – a policy to sustain products when their market prices are too low. Storage is organized by the European Commission, which can buy and store cereals when market prices are below a certain threshold. Farmers are free to apply for this measure.

The scheme is very detailed because of the necessary certification of actors and procedures. Each EU member state can adapt the policy to its internal situation; storage agencies – cooperatives or private companies – must be formally acknowledged, and payments must be audited. The European Commission decides when to sell the paid-for and stored cereals; it usually sells when prices are higher, and in certain years such activity has provided a net income for the EU, in that the costs of intervention and storage have been lower than the earnings from subsequent cereals sales (European Commission 2011: 11). Moreover:

> some quantities are released as food aid for the Most Deprived Programme (2.79 [million tonnes] in 2011 and 162,000 [tonnes] in 2012). Intervention stocks can also be exported to third countries through export tenders. In recent years the Commission has been reluctant to use this possibility, however, as it might be considered as involving export subsidies.
>
> (European Commission 2011: 4)

In any case, it is important to emphasize the residual nature of price interventions and storage activities, along with the determination of the EU to leave regulation of cereal trading to the (global) market as much as possible. This tendency emerges in three ways: (1) automatic buying-in applies only to common wheat, while for the other cereals the intervention is discretionary; (2) for most subsided cereal trades, the instrument of tender must be used, and (3) import duties and export subsidies are heavily reduced, at least compared to the EU's past measures. There are some exceptions to a fully free market: cereal imports or exports require licences, there is a differentiation of duties according to the quality of cereals, and the European Commission's discretionary interventions can still be activated in the case of high turbulence on world markets. Only the last point – moderation of market instability – indicates a certain concern for food security.

In conclusion, public cereals storage is part of two long-term processes. On the one hand, there is attention to the revenues of farmers; this has been a constant of the Common Agricultural Policy from the outset. On the other hand, there is a drive to include cereals in a global market while minimizing distortions. The EU structures devoted to agriculture must behave as a moderator with the task of maintaining the market's borders in the face of price swings. Storage thus becomes a symbol of this inhibited style of action. Paolo De Castro et al. (2013) and the Groupe de Bruges (2012) maintain that it is too a modest policy in regard to security; Valentin Zahrnt (2011) thinks exactly the contrary, while there are a variety of intermediate positions (Manders 2011; Crola 2012).

The subtitle of Zahrnt's publication – *Facts against Fears* – encapsulates the argument of scholars who are against a robust public storage policy; they observe no great food crisis in Europe in recent years. Moreover, in every part of the world, food storage has:

> three potential adverse consequences: (1) large subsidies may be required to pay for the storage costs, (2) grain quality may deteriorate and can pose health risks to the poor households who receive these grains (or the children who are fed in school), and (3) when stock gets older, generally following good harvests, governments have to sell the grain to open markets, which can distort markets and adversely affect private sector incentives.
> (Rashid and Lemma 2011: 6)

Whatever the reasons, being either in favour of or against storage are too radical as alternatives; the individual country's food balance must be considered:

> The debate is often posed as a choice between trade and stocks, but this is misleading since the two strategies can and should be complementary. Countries need to achieve a food balance. In general terms, food importing countries will need to rely on a mixture of variable import tariffs and export taxes, together with a food security stock. The precise nature of the balance will depend on the country's normal food balance, its grain staple, transport costs.
> (Gilbert 2011: 26)

Gilbert's considerations provide an explanation as to why the results of the EU's food storage policy are very weak. The regulation applies especially at the national level, where intermediation agencies and farmers' associations play an important role. The agricultural world in general, and the storage sector in particular, have a striking capacity for aggregation and membership. Despite policies intended to liberalize exchanges, the primary sector is still a thick social network deeply rooted at local and national level. Italian cereal storage centres are good examples of this situation.

Italy imports about 50 per cent of its cereals, with some differences for individual varieties (for example, it is less dependent on maize); but it is a large exporter of milled products, animal feed and, of course, pasta. This specific import/export regime leads to the large-scale movement and storage of cereals. Certified store centres for internally produced cereals number around 1,200 (ISMEA 2014). According to the Italian association of agriculture products traders Compag, the capacity of private companies to store cereals and soya exceeds that of cooperatives or consortia, even in Italian regions with strong associative traditions:

> According to a survey conducted in 2007 in Emilia-Romagna by Assincer, private storage centres have a storage capacity of 130 if considered 100 the capacity of cooperative centres; they represent 47 per cent of the entire regional storage capacity (private storing companies, farmers' cooperatives, mills, and animal feed manufacturers), which is equal to 2,650,311 tonnes (35 per cent coops, 11 per cent mills, 7 per cent feed manufacturers and farms) compared with a 2007 production of 2,611,156 tonnes. The average storing capacity of each centre, including all private companies and cooperatives, is equal to 10,000 tonnes.
>
> (Ticchiati 2008: 12; my translation)

This quotation shows that the Emilia-Romagna region is almost self-sufficient, and that the storage capacity of the food transformation industry is residual.[3] Storage centres are increasingly independent third parties with respect to farmers and the food industry. This confirms the above-mentioned self-referential tendency of storage practices; but, at the same time, it illustrates the problems: fragmentation of structures and, overall, few possibilities of differentiating between stocks because the storage plants are too small. Differentiation is the only strategy still in the hands of cereal producers since competition on production costs was lost in the globalized market. Italian as well as European producers of cereals are unable to compete with North American and Russian/Ukrainian and Australian farmers in terms of production costs. In terms of yield, they may be still competitive, but with some limits due to the protection of biodiversity and long-term land fertility.

Differentiation of lots according to *commercial* (proteins, specific weight, moisture), *technological* (gluten, colour, P/L, W and falling number) and

---

3 According to a more recent survey conducted at national level in Italy (ISMEA 2014), in Emilia-Romagna there are 177 storage centres and a smaller capacity (1,791,938 tonnes), although this is still the highest among the Italian regions (16 per cent of the total). There are only indirect indications of the independence of the storage centres from both the agro-industry and farmers: cooperative storage centres, weakly controlled by farmers, represent only 26 per cent of capacity, while storage centres with the capacity to mill or to prepare animal feed, presumably controlled by the agro-industry, are a small minority.

healthiness parameters is the objective of specific policies.[4] But numerous difficulties arise, most notably: (a) the problem of monitoring qualitative differences, a complex process due to the fragmentation of farmers and storage centres; (b) the food processing industry and consumers – the intermediate and final markets for cereals – do not recognize better qualities and higher prices, and in general there is scant public recognition of the above-mentioned quality parameters, and (c) farmers themselves are accused of behaving badly because they send the worst products to the association's storage centre (the best products are sold privately), and they store cereals in such centres for merely speculative purposes.

These issues have been discussed at length by economists and policy-makers in search of the right balance between trade and storage (see Gilbert 2011). But they have a social dimension as well: the capacity of each actor to cooperate to arrive at a better common result. There is a problem regarding trust in and loyalty to the supply chain – exemplified by the above-mentioned supply of inferior crops to associate structures – and another concerning information asymmetries in the cereals market. Storage centres play a role in the effort to reduce information asymmetries and deceitful behaviour; it is a strategy that can be conceptualized by applying *game theory* to cereals storage (Bowles 2004: ch. 11). The storage centres are asked to avoid the typical prisoner's dilemma by assuming the specialized role of monitoring the quality of cereals and giving valuable information to farmers and the food industry. The storage centres have sometimes undertaken the roles of previous and successive phases in the supply chain by respectively selling seeds to farmers or processing the cereals internally. They sometimes create agencies specialized in the management of supply chain exchanges as substitutes for the market (see Box 2.1).

Box 2.1 Cooperative Terremerse, Ravenna, Italy

Established in 1991, Cooperative Terremerse is the result of a merger of various cooperatives arising from the 'Cooperativa Servizi a Coloni' founded a hundred years ago, on 30 April 1911, to provide services to settlers (after land reclamation), small owners and hired hands at Massalombarda, a village between Ravenna and Bologna. Over time, many cooperatives operating in different fields and sectors have channelled their resources and knowledge into Terremerse, including providing agricultural products, cereals, fruit and vegetables, and processed meat.

Today Terremerse is a consolidated agrifood processing company, with a turnover of 157 million euros, able to distribute services and assets to its 5,271 members (in addition to more than 880 investing partners). Terremerse offers assistance to its members to trade products on the market and also through single crop agreements.

---

4 Parameters adopted by the Rete Qualità Cereali project promoted by the Italian Ministry of Agriculture Food and Forest Policies. The technological criteria are set out at http://www.theartisan.net/flour_criteria_judging.htm (accessed 17 September 2015). Policies are implemented by the Coordinamento Cereali, a conglomerate of most of the farmer and cooperative organizations (Sgrulletta et al. 2013).

The aim is to create sustainable quality in the food supply chain for consumers, partners, customers and collaborators, with full respect for people and the environment. Terremerse operates in the national and international markets, and it is a leading company for the provision of cereal proteins, fruit and vegetables, agricultural products, farm machinery and equipment, as well as meat.

The cooperative has adopted an innovative horizontal silo bag system for grain storage. This system requires a larger surface area, but it ensures lower plant management costs and a natural conservation of the products stored using CO_2. The system allows better differentiation among the products stored, and it allows recycling of the soil temporarily dedicated to the silo bags.

Source: http://www.terremerse.it/jdownloads/Altro/TER_Brochure_web_en.pdf (accessed 10 September 2015).

Storage, therefore, is the basis of coordinated activities. The storage location can be a meeting point for actors with different needs and mentalities. Evidently, this depends on the concrete tools the managers adopt: appropriate and rapid instruments to measure cereal qualities, internal compartments for the storage of differentiated stocks of cereals, and healthy measures for the long-term conservation of cereals (Fleurat-Lessard 2002). The long-term preservation of cereals in good condition is important, not in order to speculate on prices, but to regularly provide a wide range of varieties to the food processing industry in response to frequent complaints from the food industry and major retailers, which are driven to import foodstuffs from abroad because the farmers' dispersed system is unable to provide the right quantity and quality at the right moment.

These issues are similar to those in less-developed countries, where the fragmentation of food producers is very high and the concentration of the food industry and retailers is even greater than in the most-developed countries. Farmers' cooperation to enable tolerance of differences in the quality of crops is also a good criterion for analysis in countries with problems of food shortage and undernourishment. Individual farmers are of course very weak, but cooperation fails – as has happened many times in the Italian countryside – when the differences among farmers' products are not appreciated. Indiscriminate storage is emblematic of this mass deterioration of food provision.

It may seem strange to talk of *quality* and *differentiation* in countries where there is not a minimum level of food provision and food shortages are endemic. In some cases, only large-scale external assistance (so-called 'humanitarian stocks') can prevent malnutrition or, worse, starvation. But wherever agriculture is organized to any extent – of which common storage structures may be the signal – a quality control service is considered necessary. The problem lies in sharing norms and standards in order:

> to avoid disputes about the quality of produce in both operations to supply the reserve and the distribution of foodstuffs in a crisis. Initially, a broader

diversification of foodstuffs can be initiated at the level of national stocks, to later be extended to the regional level.

(Economic Community of West African States 2012: 8)

A certain degree of differentiation is thus envisaged and encouraged; but the extent to which it is practised and rewarded is not clear. In any case, it is a first step towards *sovereignty*, understood here in a very specific sense as the capacity to manage cooperation and differentiation simultaneously in local and poor economies.

2.3 Healthy food: storage as preservation

Food safety is a puzzling concept. In certain respects, it is an aspect of food security, because availability of pathogenic food indicates an inability to exercise control over it. In this regard, safety is one dimension of the more general theme of certainty in food provision. In the French and Italian languages, there is only one term for safety and security – *securité, sicurezza*. However, safety emerges as a specific category in two cases: (a) when food is available in such great quantity that quality becomes the issue, for example if apples are abundant and cheap, but polluted with pesticides, and (b) when the food processing industry is global. In the latter case, the risk of transmitting foodborne disease throughout the world becomes a concern at the top of the international agenda. Also, agri-food companies have increased their attention to providing healthy food. Their success depends on images of goodness and pleasure disseminated through the mass media. If an isolated incident occurs – say, the dramatic poisoning of a consumer – trust in their products plummets, and economic and reputation losses for private bodies are enormous. Panic situations are also feared by governments, which may decide, even without certainty about the real causes, to withdraw the product from the market. That indicates how safety and security are intertwined and entrenched in the public sphere.

Moreover, in our framework, food safety is a special category emerging from the cross-referencing of (self-)control and purposive rationality (see Figure 2.1). Safe food simultaneously provides an opportunity for exercising control by every organization and a source of rationalization. Undoubtedly, safe storage allows better control over access to food and easier planning of its consumption. Furthermore, families and factories, communities and governments want to keep their members healthy through efficient organization of food provision. This aim is not a matter of choice; it is a necessity to which the best means of achieving it must be devoted. In fact, purposive rationality entails focusing less on the aim and more on the best way to achieve it. In economic terms, we say that individual preferences are taken for granted (Palermo 2003: 7).

At first glance, safety appears to be an issue of the same type: healthy food is taken for granted, not only at the individual, but also at the community level. Unlike single preferences in the economic realm, access to safe food is a right of every citizen and an obligation for the authorities. It is a public goal. But the basic mechanism is the same: achieving the goal in the most efficient way possible. Storage is an important part of this process. It has been necessary to store food since humankind first engaged in agriculture, but the problem is that its preservation became progressively more complex and risky. Cheap and practical ways to preserve foods in appropriate forms for long periods had to be discovered.

There are numerous food preservation techniques, and they depend closely on the characteristics of the context. The following list conveys the wide range of such techniques:

> Drying, Pasteurization, Refrigeration, Freezing, Vacuum packing, Salting, Sugar use, Smoking, (Use of) Artificial food additives, Pickling, Lye, Canning and bottling, Jellying, Jugging, Irradiation, Pulsed electric field electroporation, Modified atmosphere, Nonthermal plasma, High-pressure food preservation, Burial in the ground, Controlled use of micro-organisms, Biopreservation, Hurdle technology.

It is taken from the *Wikipedia* entry 'food preservation'.[5] It includes twenty-three methods, which is a large number – certainly more than most might imagine, even if they are quite expert in cookery. Reading the description of each technique raises important issues for food storage and safety. First, some techniques are very old and used more or less in the same way today (drying); others are very old, but are re-used with major variations (modified atmosphere). Second, some techniques require a great deal of sophisticated equipment (irradiation, pulsed electric field electroporation), and they can be used only in special premises by experts. Third, some techniques are controversial (irradiation); they are accepted by the food and health authorities, but contested by ecological organizations. Fourth, each technique combines in a particular way two objectives: maintenance for a long period of food's original features like texture, taste and nutrients, and control of toxic substances or organisms.

Food storage thus confirms its multidimensional nature, cutting across tradition and innovation, lay and expert knowledge, the small and large scale, ubiquitous and super-localized practices. There is, however, an old distinction of great importance: that between fresh and processed food. This distinction recalls another basic dichotomy used to classify the human venture: the distinction between raw and cooked has been well known since the work of the anthropologist Lévi-Strauss (1970). Cooking food, besides marking a distance of culture from nature, was a way to make food more digestible and – the feature of interest here – a way to preserve it. Transformations of food by fire, heat or salt had the primary function of preserving it – in other words, keeping it safe for a longer period. Modern raw food diet followers would disagree. In any case, storage is an instrumental process whose purpose is to preserve edible food. Another primordial distinction thus emerges: that between edible and healthy food (Nordström et al. 2013). 'Edible'

---

5 https://en.wikipedia.org/wiki/Food_preservation (accessed 21 January 2015).

means that food can be eaten with at least a minimal contribution to the body's nutrition, that the taste is agreeable, and that the body's enzymes are able to process it. Food safety is defined by default as absence of harm to the body, and this relates to issues of quantity. Sugar and sweeteners are perfectly edible, but if eaten in great quantity they provoke well-known health problems.

Returning to the fresh/processed food distinction, to be noted is that this is a continuum rather than a sharp separation (International Food Information Council Foundation 2010). Very few comestibles are consumed without any treatment except collection in the field, simple washing with water, and rapid transport to places of consumption. They are essentially vegetables and fruits, which in any case can be processed, as testified by the growing importance of convenience food.[6] Most food consumed is processed with more sophisticated procedures, of two types: those that change the form of the food, for example from wheat to flour through milling, and those that combine different materials to form a new product, such as bread making. Indeed, some procedures radically change the nature of the food: for example, in the fermentation of grapes, most of the sugar becomes alcohol.

The *risk analysis* that consumers, experts and authorities perform along the fresh/processed continuum is noteworthy. With treatment, food acquires durability, but some of its qualities are lost or depleted. Furthermore, the same treatment can have secondary effects. The choice of where along the continuum to locate a specific food is plural and sequential at the same time. It is plural because different actors and techniques intervene; it is sequential because the transformation takes place in distinct stages or along 'technological paths' (Hand and Shove 2007: 87).

The example of bread highlights how risk analysis and production chains combine. We mentioned above the changes to milling that occurred in the nineteenth century. The manufacture of white bread was promoted by the introduction of roller mills and by the desire of the upper classes to consume a distinct item. In recent years, a movement for the reintroduction of brown bread has begun, and the modularization of mill-bakery technology makes this possible. Thus, there is the possibility for consumers to exercise multiple choices within the bread chain: whether to make it completely at home, whether or not to use flour that includes roughage, and whether to buy the flour directly from a local mill. According to the procedure chosen, different actors intervene. With the existence of many actors and procedures, compliance with food safety prescriptions becomes more difficult.

Research on food safety shows that foodborne diseases occur much more frequently in the final stage of food processing, the one that concerns the household and final consumption (Redmond and Griffith 2003). This is an important finding because it shows that industrialized food processing is safer than

---

6 Convenience food, also called 'tertiary processed food', is commercially prepared food designed for ease of consumption (in Italian, *cibo pronto*). In particular, fresh vegetables are provided already cleaned, cut and packaged.

processing in the home. This is contrary to the perceptions of public opinion concerned about dangerous procedures adopted by the agrifood industry. The standardization of procedures is cited as an explanation of the greater safety of industrialized food. In highly automated procedures, the sources of contamination are reduced: the workers employed are adequately equipped and trained, and the environment is sterile. At home, people generally process food in very crude ways, without the capacity to control contaminants. Factories can reproduce the conditions in experimental laboratories, where external factors are kept under control. Expertise and aseptic fields are therefore crucial for risk control in the food chain.

The scientific contribution to food safety is very wide and manifold. First, scientific knowledge enables systematic observation of processes, thereby establishing with statistical models possible correlations between food handling and diseases. Second, scientific knowledge makes it possible to show aspects usually too small to be seen by lay people: the presence of bacteria in food can be discovered only with powerful microscopes. Third, scientific knowledge furnishes tested indicators of contamination or deterioration in food and in food processing – indicators with two features: rapid signal and low cost.

All these objective advantages are under severe scrutiny by 'post-normal' science scholars (Funtowicz and Ravetz 1993). Nevertheless, the high knowledge content helps us to understand why food safety is supervised to such an extent by experts. Here the word 'expert' denotes a figure with three qualities: long experience in the specific field, a qualification obtained from a scientific authority, and involvement in the decision-making process. According to the theory of naturalistic decision-making (NDM), expertise concerns 'how experienced individuals or groups of people make decisions in dynamic, uncertain, fast-paced environments' (Krusemark and Block 2011: 74). Decision-making is what distinguishes an enrolled expert from a pure scientist. Experts prearrange the options for other people's choices, or even make the choices themselves. This implies distinguishing the scientific from the political phase, or in the terms of Millstone (2009: 629), 'differentiating "risk assessment" from "risk management"'. In regard to food safety evaluation, Millstone maintains that 'the alleged separation of science from all policy considerations is illusory. In each jurisdiction some "risk management" policy issues are being decided by scientific advisory bodies, typically but not invariably acting as "risk assessors"' (2009: 631).

According to Millstone, there is another crucial aspect of food safety expertise: 'the responsibility for risk managers to provide their risk assessors with up-stream guidance' – a task he calls Risk Assessment Policy (RAP). This should include a plurality of evaluation styles (substantive, procedural and interpretive) in order to achieve an ideal co-evolutionary situation in which RAP issues are:

> decided *in advance* of risk assessment, *in consultation with* risk assessors and *all interested parties*, in processes that are *systematic, complete, unbiased and transparent*[;] then the conditions may be in place for this set of

science-based risk policy decision-making processes to become both scientifically and democratically legitimate.

(Millstone 2009: 634–635; original emphasis)

According to this model, experts must prepare with other actors a decision within a public procedure to establish whether or not a food preservation substance or technology is safe.

Despite all the procedural assurances, criticism is made of expertise applied to food preservation safety. It is based on three considerations:

- Evaluation methods are often misleading because they are de-contextualized (Paul 2008). They are too simple in a variety of environments with the possibilities of compensation or escalation among local processes. There is a well-known claim among some cheese producers that the adoption of strict hygiene measures to make safe cheese reduces the flavour, or even blocks the curdling process.
- Evaluation methods are formulated by experts who are not impartial because they are under the influence of the food industry (Young and Quinn 2015). Food authorities cannot know everything about food. Consequently, they enrol external experts who are often professionally linked to the firms whose procedures are to be evaluated. The entire evaluation system is based on references to publications, raising the same criticism (Batista and Oliveira 2009: 284).
- Evaluation methods are biased toward preventing 'false positives' (see Pellizzoni 2010) – for example, rejecting a food preservative because faulty experiments show a possible connection with the incidence of cancer among consumers. In other words, the experts working for public agencies are more afraid of blocking an authorization (the precautionary principle) than of giving permission for a harmful substance (the 'false negative' error).

The first criticism arises from the ecological perspective: the approach to hygiene of mainstream experts is based on a reduced number of variables and environments, those typical of the areas where there have historically been more food safety experiments. This criticism also assumes an ideological flavour, leading to the idea that food preservation norms are ethnocentric, biased toward a restricted vision of the world (MacCoun 1998). This perspective is connected to the second criticism: the weak independence of experts. This is a longstanding criticism of the scientific world, first levelled by Robert Merton and then by Marxist scholars. The result is clear: the dominance of expertise combines with industrial or capitalist dominance to create a sharp socio-economic divide in the world (Cutler 2010). The third criticism is less easily framed into the previous dichotomies, but there is a connection: in very broad terms, food authorities are considered too submissive to science, expertise and high-tech industry. There is a complex of institutions, companies and academic experts (a 'triple helix') that is generally favourable to every proposal made by large, rich and specialized organizations, with a reverence for supposedly universal values of modernity based on innovation and economic success – and, indeed, expertise. This world view is important because it affects the attitudes of controllers and lawmakers in the field of food preservation.

The case of genetically modified organisms (GMOs) can be framed in terms of these opposite views: one favourable to change and ready to accept risks, the other more pessimistic and invoking the precautionary principle. Expertise is biased by strong pre-scientific assumptions (Jasanoff 1987), and food processing evaluation is no exception to this rule. These assumptions are consolidated in scientific cultures. This appears evident in the different attitudes to food risk in America and in Europe: 'EU rules are more risk averse ("precautionary") than those of its trading partners, including the United States': this is the key phrase in the blurb for a book by Ansell and Vogel (2006).

The answer to the complexity of food safety is extensive use of transparent authorization procedures. Let us consider how the European Food Safety Authority (EFSA) normally works on a wide range of issues. For the evaluation of packaging – a crucial component of food preservation – there are two panels: Additives and Nutrient Sources Added to Food (ANS) and Food Contact Materials, Enzymes, Flavourings and Processing Aids (CEF), supported by the Food Ingredients and Packaging Unit (FIP). All meetings are recorded, and the minutes are published on the EFSA website. At every meeting, panel members make a declaration of any conflicts of interest. Thus, formally, the rules of transparency, peer community review and impartiality are rigorously observed.

Of course, the sources of collusion with the packaging industry can be very subtle: they concern not only the chances of receiving large grants for research, but also mental affinity between experts and industry managers. In the specific case of packaging, striking a balance between safety requirements and the prospects of recycling the packaging materials is not easy, and must respond to different expertise philosophies. The former concern is focused on human health, the latter on saving energy or reducing environmental pollution. For example, a dilemma emerges when analysing the choice between using Tetra Paks or glass for liquid conservation. The former is a package under the full control of industry, and therefore presumably safer; the latter leaves more room for intervention within the storage chain, and therefore is less safe, but more easily re-used or recycled.

In conclusion, we have identified a dual linkage between storage and safety (see Figure 2.2). There is a circularity between the two poles that creates an escalation of regulation. Food storage is progressively modularized in order to respond to the imperative of production and consumption differentiation. This entails a similar process in the techniques of food preservation, which in turn entails new risks and the need for new evaluations.

The responses to this escalation give rise to a wide range of attitudes to food: the search for a simplification which imitates natural processes (organic food and preservatives), and the search for new artificial and safer materials to be used for packaging and as additives. In the middle are 'artificially

40 *Food processing and consumption*

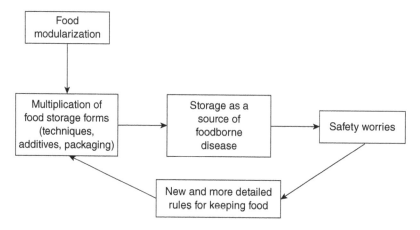

Figure 2.2 Circularity between development of food storage techniques and safety measures

reproduced natural techniques' like refrigeration/freezing and drying/heating. The former seems the safest,[7] but it is highly demanding in terms of secondary energy. In fact, it needs a regular supply of electricity. This raises new issues: the efficiency of food storage practices, and their environmental impact. Both will be discussed in the next section.

2.4 Organizations opposed to food waste

Storage has been framed as a way to reduce food waste and spoilage. Storing food is conceived as a rational activity intended to improve the efficiency of provisions. All the food chain actors are aware that wasting food is a shame and an unsustainable cost in a period of scarcity. According to a rational actor model, a genuine effort to reduce waste is expected at every stage in the chain. We can divide the topic into three parts:

- the effort by industry to provide the best food storage items for saving energy, packaging materials, money and the food itself;
- the effort by some organizations, usually non-profit, to save food at the final stage, from retailers to consumers;
- the devices used by consumers to reduce food waste and spoilage.

Food storage serves several purposes. As seen above, it is a strategy used by agricultural operators to speculate on prices; it is a precaution taken by governments,

---

7 Rapid freezing to about -18 °C gives assurance of halting the growth of bacteria (Potter and Hotchkiss 1998: 165), while there is still the risk of botulism with canned food.

especially in troubled periods, to ensure at least minimal food security, and it is a device employed by rational consumers to optimize the preparation of meals. Last but not least, food storage is also an industry – a large and powerful industry involving packaging, chemical additives, machinery construction and maintenance. Because food storage has become so varied and complex, industry has responded by working in two main directions: research and development (R&D) and public relations. R&D meets the need for better-performing products, while the management of external relations (advertising, lobbying, social responsibility and so on) is necessary to deal with the growth in public regulation of food safety.

The packaging industry[8] is in a situation quite similar to that of the overall manufacturing sector. It regularly increased its total revenues, at least until the world crisis of 2008. It has developed in particular in Western countries, plus Japan and China. It follows the general trends of consumption. It therefore accounts for the development of smaller portions, convenience food and 'growing requirements for brand enhancement/differentiation in an increasingly competitive environment; [though] health awareness was regarded as the single most important driver to growth in the packaging industry' (World Packaging Organisation 2008: 5). This is the self-representation of the industrial sector. Externally, the processing of packages after their use for food is seen as the crucial issue, of concern not only the industry, but also the final consumers and waste management authorities.

The strength of the packaging industry relies on demonstrating that it is able to provide consumers and retailers with healthy and palatable food in practical ways to suit any circumstance. It is able to do so because of the great flexibility and variety of products and services. Inevitably, there is competition within the sector, especially between the two historically most industrialized methods of food preservation: canning and cold chain. We can compare them and other, newer methods according to three criteria:

- maintenance of original food features (structure, taste, nutrients);
- energy consumption/environmental impact;
- practical usability.

For the first criterion, the comparison is markedly in favour of the cold chain method. Chilled food lasts for long periods, maintaining its main features almost intact. We have already mentioned its superiority in terms of safety. The competition is more balanced not for freezing, but for refrigeration, in particular for the food's life span; food in a refrigerator does not last more than two weeks unless it is sealed. Canned food, by contrast, can be consumed even after years of storage. An important advantage of the cold chain concerns the external features of food,

---

8 According to the World Packaging Organisation (2008: 3), 'the global packaging industry turned over around $485 billion in 2004 with packaging container sales of almost $460 billion and machinery sales of around $25 billion'.

which remain almost intact. As regards the internal ones, matters are somewhat more complex because for some fruits and vegetables rich in water, freezing tends to break down the cellular structure, modifying both appearance and solidity.

For the second set of criteria, the comparison is more favourable to canned food. According to the Netherlands Organisation for Applied Scientific Research (TNO), which conducted research for the Association of European Producers of Steel for Packaging on eco-efficiency (environmental impact plus *shadow price*[9]):

> canned carrots, fresh peeled carrots and frozen carrots sold in a bag have a comparable, and slightly above average eco-efficiency. Fresh bunched carrots perform, due to their below average environmental impact, the best when considering the Dutch products. The least performing alternatives are frozen carrots sold in a carton, carrots in a pouch and carrots sold in Tetra Recart carton laminate. When considering the Dutch market products, the canned carrots are the best performing products.
>
> (Lighart, Ansems and Jetten 2005: 3)

This research, even if limited to one product, demonstrates two important things. First, food storage is very often a coupling of conservation techniques and special containers; their variable combination can yield opposite results in terms of eco-efficiency. The case of carrots preserved in pouches provides an example. This type of packaging – consisting of multiple layers of flexible laminate – achieves a good rating for environmental impact, but its shadow price is quite high. Hence, the final eco-efficiency result is poor – the poorest according to the research. Second, canned carrots achieve a good result also because in the Netherlands there is a public–private agreement for recycling the steel used for containers. This reduces the shadow price of this kind of package. The agreement does not guarantee that the final consumers will recycle the steel containers, but the commitment of producers to take back a quota of the used containers reduces the shadow price of canned food. As we know, the accounting method affects the results:

> In terms of energy, the two storing techniques – freezing and canning – seem very different: generally, frozen food absorbs much more energy than canned food. The comparison with refrigerated food is more balanced: 'the most energy effective method for product delivery is canned ready meals (1136 kcal/lb) followed by bulk refrigerated (1152 kcal/lb), canned fruits &

---

9 'Opportunity cost of an activity or project to a society, computed where the actual price is not known or, if known, does not reflect the real sacrifice made' (http://www.businessdictionary.com/definition/shadow-price.html#ixzz2rjQ6iqfb; accessed 17 September 2015); in our case, shadow price is the cost of reducing CO_2 emissions with the best available technology for producing a kilogram of carrots. The TNO study calculated that the shadow price of packaging is about 1/25 of the consumer cost of the carrots.

vegetables (1607 kcal/lb), refrigerated portions (1692 kcal/lb), frozen boxed products (2250 kcal/lb) and frozen bagged products (2406 kcal/lb).

(Ritchie 2012: 29)

These results must be balanced with considerations of how long the packaged food is kept in storage. According to David and Marcia Pimental (2008), when comparing the energy (kcal or watt) required for processing and packaging the food according to the two methods, the difference is very low (see Table 2.1). Hence, the real difference concerns the use: if the intention is to keep the food for emergencies, and therefore for a long, undefined period, canning is certainly less energy-demanding than freezing.

The third set of criteria concerns practical use. In this case, the comparison is less easy because the two preservation methods depend greatly on the circumstances and on subjective preferences. The strength of canning is that it makes it possible to store food at room temperature for a very long period without a high risk of foodborne disease. It also has the advantage of easy mobility, because canned food is not particularly heavy. It is therefore ideal for highly mobile people, such as soldiers and explorers. Moreover, the canned food is already cooked or peeled, so it is ready to eat.

The cold chain requires two important devices: a constant source of power, and a heat pump connected to a well-insulated container. It is evident that, with a billion people in the world without electricity available at home, the practical use of cold is greatly reduced. In this regard, experiments to bring electricity to remote rural areas may be helpful, given that refrigerators have become relatively cheap. At the same time, attempts to recreate the old methods of keeping food cold (or at least cool) are noteworthy. They concern root cellars (purpose-built holes in the ground) or simpler clay pots.[10] Of course, these are techniques for refrigeration, not for freezing; consequently, food stored in this way cannot last more than a season. Furthermore, root cellars need an open space near the house. In the case of 'passive' off-grid houses (see Chapter 3.2), the revival of the old store room

Table 2.1 Example of energy (kcal) used to store 455 g of corn

	Packaging	*Processing*	*Total*
Freezing	722	1,550	2,272
Canning	1,006	1,300	2,306

Source: Author's elaboration of Pimental and Pimental (2008: 251–252).

10 Zeer, or clay pot, refrigeration keeps food cool (in fact, icy cold) without electricity by using evaporative cooling. Essentially, it consists of a porous outer earthenware pot, lined with wet sand, containing an inner pot (which can be glazed to prevent penetration by liquids) within which the food is placed. The evaporation of the outer liquid draws heat from the inner pot; http://www.greatnorthernprepper.com/food-prep/food-refrigeration-techniques (accessed 8 May 2015).

is possible in urban contexts. Improvements in insulating materials have been so great that a passive cellar within the house is practicable.

In the same vein, the canning of food can be replaced by old methods of preservation like chemical pickling, whereby possible toxic agents are removed:

> the food is placed in an edible liquid that inhibits or kills bacteria and other micro-organisms. There are many pickling agents include brine (high in salt), vinegar, alcohol, and vegetable oil (particularly olive oil). Many chemical pickling processes also involve heating or boiling so that the food being preserved becomes saturated with the pickling agent.
>
> (McKeen 2012: 7)

As mentioned above, glass containers are very practical because they are manageable at lower levels of the food chain; their use for final consumers is mainly limited to vegetables and fruits, but extension to fish and meat is possible through repeated sterilization using hot water. For some fish, like sardines and cod, there are again old domestic practices that can be revived after the cannery industry has discarded them. The ease of food storage at consumer level depends greatly on the potentialities of old practices; such possibilities are in turn linked to the availability of fresh and dry rooms in the house and to the willingness to learn or develop retro-innovations. Generally, there is no knowledge exchange between household practices of food preservation and the packaging and storage industry. Each is a closed circuit with its own technological path. However, mediation occurs for specific materials, like plastic. Plastic bags, especially when vacuum-sealed, are highly promising in the field of food preservation. The device for creating the vacuum is simple and available at all the chain levels.

The real problem is recycling the hundreds of types of plastic used for food packaging. This is not just a problem for certain actors in the chain, but for society as a whole – a typical commons. Evaluation of the practical aspects of packaging cannot avoid this problem, according to the principle of 'shared responsibility' of producers and consumers for waste management (Coggins 2001). According to the US trade association for the metal can manufacturing industry and its suppliers (Can Manufacturers Institute 2013), there is a precise hierarchy in the recycling of solid wastes: at the bottom are flexible materials or plastics, then cartons and glass, and finally metal containers. The best are food steel cans, with a 71 per cent recycling rate. The Natural Resources Defense Council confirms this ranking, with the addition of paper and paperboard, which rank top among recycled packaging (MacKerron 2015: 10–11). In any case, the overall US capacity for recycling packaging is lagging far behind that of western European countries.

The term 'recycling' introduces the central theme of this section: food waste. This issue is typical of societies with food abundance and class disparities. The rational use of entire portions of food was not a problem for the poor strata of pre-modern society, which developed dedicated practices for food recovery. In terms of pig husbandry, the motto 'everything but the squeal' is an example of this. Rural people were able to eat or use all parts of the animal. That fullness of use has disappeared since the advent of abundant raw materials and energy. Industrialization alone cannot be accused of wasting food, because one presumes that every agrifood factory seeks to save as much as possible; the same applies to consumers, who also try to rationalize their material lives. The causes are the decreasing value ascribed to food, along with poor connections within the agrifood distribution chain.

As people become more affluent, they attribute less monetary value to food; this is evident in the proportion of the household budget spent on food in developed countries and among high social strata, according to the United nations Food and Agriculture Organization (FAO 2011: 14). If food is progressively cheaper as a proportion of the family income, then wasting some of it is of less importance. Poor people, even if they have diets rich in high-calorie food, pay more attention to avoiding waste: 'The emphasis on maximum calories and least waste and spoilage is another characteristic of poverty' (Darmon and Drewnowski 2008: 1,111). It is sad to discover that 'because trying a new food represents a risk of waste [. . .], diets of low-income households are often monotonous' (Darmon and Drewnowski 2008: 1,111). Past experience of poverty gives rise to a similar attitude: older people who have experienced scarcity store more and tend to consume expired food (Terpstra et al. 2005).

There is a second mechanism linked to the deficiencies in food distribution. Most food is wasted between the retailing and consumption phases because of a bad match between supply and demand. The intervening factor is food waste as a result of expiry dates. The same problems noted above in terms of food safety apply to use-by dates: (1) the life span of a specific food is difficult to evaluate, so the threshold is arbitrary (cognitive problem), (2) the packaging industry is more interested in the industrialization of processes than in ensuring the best preservation of food (goals heterogenesis), and (3) local storage conditions vary greatly compared with standards established by food safety authorities (de-contextualized norms). Overall, the distribution system is too rigid compared with the relative freedom of the final consumer's decision on whether or not to comply with the expiry date.

The main problem is for retailers and restaurants, which must comply with the expiration rules established for food served in public places. Rules governing not only use, but also conditions of food preservation (hygiene, dry storage and so on) must be respected. A conscientious retailer or bar manager will have to throw food away some hours before its expiry date. There are many reasons for the failure to sell food before its expiry date; it is not always due to managerial errors in forecasting sales. In any case, it is clear that a free market mechanism does not work well with the imposition of an external rule such as food expiry dates, because the demand is unpredictable and the supply is too rigid. It fails every time food is wasted. For this reason, corrections have been made to the functioning of markets.

One typical correction on the supply side is overbidding: the sale of items that are near expiry at a lower price, often below cost price, promoted through buzz or viral or word-of-mouth marketing. The organized response on the demand side is 'consumer deal hunting' – links among consumers who inform each other which shops are charging the best prices for items. In the USA and Europe, such consumer organizations have become a mass phenomenon: coupon groups have millions of members. In this case, they have already been captured by clever companies (Edelman, Jaffe and Kominers 2014), which once again demonstrates the extraordinary capacity of market mediators to intervene between demand and supply.

However, the most interesting phenomenon is the advent of non-profit organizations specializing in the collection of near-expiry food and its distribution to community centres helping poor people. They are generally termed food banks – in France, *banques alimentaires*, with their network's headquarters based in Paris. In Italy there are two national schemes, one religiously inspired (Banco Alimentare) and the other more lay in nature (Last Minute Market), both aiming to reduce the shameful waste of food by large retail centres. Similar organizations exist in every country. Many aspects are edifying, others more problematic, but in any case, food storage techniques are crucial for all of them.

Such organizations work preferentially in the middle of the above-mentioned continuum between fresh and processed food. It is difficult for them to deal with fruit and vegetables because in two or three days the fresh products are no longer edible. With access to ultramodern storage techniques like controlled atmosphere, they will probably be able to save more fresh food; but such techniques are obviously expensive and, in particular, only modestly modularized (confined to large stores). In the case of long-term usable food such as canned or frozen products, retailers are equipped with the appropriate technology to allow them to manage near-expiry foodstuffs more easily. Finally, there is a set of products – yogurt, milk, tortellini, convenience foods – that have short expiry dates, from several days to a couple of months, and are more sensitive to small variations in sales. Non-profit organizations have a particular role as social intermediaries in terms of this category of semi-processed food:

> In a study on Canadian food banks, Tarasuk and Eakin [. . .] identify two critical aspects: the expectation that beneficiaries will accept with gratitude any kind of food and, on the contrary, the fact that this kind of charitable distribution is actually a channel for the disposal of waste food, with the consequent weak nutritional quality of most provided products. The same conclusions are reached by Alexander and Smaje [. . .]; moreover, they highlight the difficulties and costs of food collection.
>
> (Orlandi 2015: 196–197; my translation)

The storage/transport combination is important as well. Such organizations need a storage space where they can sort the collected food. The usual minimal features of stores are necessary: a cellar or refrigerator, and a dry and clean environment away from heat sources and direct sunlight. The size and location of the storage structures are crucial for the efficiency of such organizations. The use of warehouses seems impossible even for organizations that are very rapid in collecting and redistributing the items. Other organizational aspects required include the presence of staff and possibly computerization of records (Eisinger 2002). In conclusion, the modularity of devices (for example, a range of transport means) and the right mix between volunteers and paid staff are key organizational factors even for small street-level charities animated by ideals of generosity.

2.5 Rituals of food storage

The symbolic value of food storage brings us back to the realm of cultures and identities. Food storage is charged with deep symbolic meanings. According to the scheme in Figure 2.1, we are not dealing with pure repetitive behaviours, but a more comprehensive rationality that includes the means–end logic and enriches it with discourses about sense-making, trust in people and God, and concern for others and the environment. Appadurai (2004: 64) makes a formidable synthesis of the framework by urging 'due attention to internal relations of cosmology and calculation among poorer people'.

We will develop the idea of wider rationality by examining the main example of food preparedness inspired by religious belief: the Church of Jesus Christ of Latter-Day Saints, which has a specific rule regarding storing food in order to be self-sufficient for at least three months. This rule has generated a large body of thought among Mormons. The eloquent online text by Cheryl Driggs, 'Food storage as a way of life', envisages three main lay or commonsense reasons for storing food. Listed in order of increasing sociological importance, they are as follows:

1   Food storage is economically rational because it reduces the number of shopping trips, and consequently the purchase of useless items. Moreover, storage is justifiable because it 'can also save time in planning and preparing meals when food storage is built around a list of tested recipes or menus, when home canned "convenience" foods are stored, and when time saving cooking methods are used such as slow cooking, pressure cooking and cooking ahead and freezing'. Is should be noted that this quotation contains all the ingredients of a rational enterprise (time saving, planning and anticipated choices).
2   It is a mode of conduct that develops self-reliance, life mastery and confidence in one's abilities: 'In actuality, food storage contributes to a better way of life and helps us be more celestial in nature. It should be taught as a principle of provident living and self-reliance, part of a way of life. Provident living is being wise, frugal, prudent, making provision for the future while attending to immediate needs.'
3   Being prepared for adverse conditions of course gives a sense of security, but not only this: when people are well prepared, they are better able to help others because they can be less emotional and have something to share. The language becomes more religiously marked: 'Food storage allows us to serve and to serve better. Elder Robert D. Hales stated "When we live providently, we can provide for ourselves and our families and also follow the Savior's example to serve and bless others." [. . .] In the pamphlet "All is Safely Gathered

In: Home Storage" The First Presidency writes "Our Heavenly Father . . . has lovingly commanded us to 'prepare every needful thing' [. . .] so that, should adversity come, we may care for ourselves and our neighbors and support bishops as they care for others." Marion G. Romney stated "Without self-reliance one cannot exercise . . . innate desires to serve. How can we give if there is nothing there? Food for the hungry cannot come from empty shelves."'[11]

The third meaning is again connoted by deep community rationality, in the sense that solidarity, reciprocal help and shared services must be carefully organized in advance. Immediate consumption must be reduced in order to store foodstuffs that will later be useful for all. It is wise not wait for a disaster, when everyone will be desperate and in need. In fact, this movement for food storage is called *preparedness* in the USA. It has given rise to a large number of guides, handbooks, lists of recipes, exchanges of information and meetings. It extends to energy and first aid devices as well.

It is important to consider the dialectic between the religious and lay meanings of food storage that arises around the Church of Latter-Day Saints. The above-listed reasons are sharable between believers and non-believers. The religious 'spark' has become a sort of secularized prophecy of preparedness, as has been said of wider universes of meaning like the Marxist ideology of progress (Löwith 1953). The transcendental interpretative register is not misused because the religious connotation of food storage is very strong, and its origin is indeed in the country of Christian sects – the USA. Self-initiated food storage has to do with the sense of future; it is based on the idea that the second coming of Jesus Christ (*parousia*) can happen at any moment and followers must be ready.

According to various passages in the Bible, the end of world, the *parousia*, will coincide with natural and human-made cataclysms. Hence, followers must be ready with carefully stored food and water. Preparedness is both spiritual and material: it concerns the intimate relation with the supernatural, and a methodical accumulation of things. It is not clear when and how Jesus will return, nor how long the period of transition will last. This belief creates a special situation in which: (a) a mystical attitude to closeness to god is developed in daily life by reading holy texts, praying, attending rituals and so on, and (b) discrete accumulation of material goods is carried out in order to maintain appropriate vigilance (alertness). Supplies necessary for survival in very precarious circumstances must be kept within reach.

We must recall that there is a contrary message in the Bible: do not worry about the future – especially by accumulating material resources – because God will provide for you, as well as, and more than, the birds that 'do not put seeds in the earth, they do not get in grain, or put it in store-houses' (Matthew 6:26), but nevertheless find the food they need. This is the providential vision of faith, which

---

11 The quotations in this list are from Helen Driggs, 'Food storage as a way of life', http://www.simplyprepared.com/food_storage_as_a_way_of_life.htm (accessed 8 August 2014).

has also been interpreted as a negation of preparedness and activism: living day by day, without planning everything, and adhering quite closely to the ideal figure of the hermit or the bohemian. This way of thinking has political implications as well: it is useless to fight to change the course of history because God is the real engine of everything. In other words, there is a temptation toward political apathy.

As theologians maintain, the Bible is full of ambivalences and paradoxes, a way of communicating that stimulates the going beyond knowledge that is taken for granted. Nevertheless, from a sociological point of view, the issue is how faiths affect concrete life. According to Durkheim (1912), beliefs in sacred things are representations of collective norms intended to reinforce the unity and efficacy of society. In contexts of high environmental danger due to hurricanes, earthquakes and floods, belief in a different future exhibits all its strength and rationality. The inclination for preparedness is also rational from a social point of view, because it is easier to deal with an emergency when everyone is well prepared, preventing looting, conflicts and mass panic, and making emergency coordination easier.

Nonetheless, the functional interpretation of religion (Horton 1960) is too narrow. It is preferable to think of a circularity or a dialectic between faith beliefs and social norms. The prescription not to eat cow meat among Hindus is not simply the sacralization of an animal useful for agriculture. Food prescriptions are used to maintain a social distinction, a common identity. They can be norms generated accidentally by local temporary 'effervescences' – Durkheim's term for moments of collective euphoria – without a clear functionality for social order (Jedlowski 2009). It is therefore more fruitful to search for further original combinations of beliefs and food storage norms than to insist on the functional role of religion. If a variable connection between religions and norms is plausible, at the same time attitudes to future events are closely conditioned by specific historical and geographical contexts. The religious symbolizations of food storage appear to be a specifically North American.

Religious references to food preparedness are not to be found in Europe; a prevention culture does exist, but it seems limited to healthcare measures like maintaining a medicine cabinet in the home. The reasons for this may be that: (a) the typical millenarist attitude of Americans is less widespread in Europe; (b) churches which emphasize readiness are not widespread in Europe, the main Christian confessions being less committed to the idea of being prepared for the end of the world, and nor has Islamic eschatology developed a similar set of beliefs/actions; (c) a relatively peaceful social context and a temperate physical climate do not facilitate sacred dramatization of preparedness, and (d) the greater secularization of religion, end-of-the-world belief included, has induced people to be less afraid of extreme futures, the end instead conceived as a placid return to the earth. In Europe, the more marked secularization of 'end time' has been translated into ecological worries and behaviours not specifically linked to religions or churches (Tucker 2006).

Another way to address the issue is to see whether there are differences in profane preparedness behaviour among European countries and try to connect these with religion affiliation. This approach was taken by Durkheim in his studies on suicide, and it is used later in this section to analyse the presence of survivalists or 'preppers' in certain countries. In more qualitative terms, the experience recounted by a former member of the Mormons, who embraced the Christian Orthodox faith, is interesting. He said:

> storing up provisions is not a bad idea, and anyone would be wise to take a cue from the Mormons in this regard. But the Orthodox Church does not obsess over 'last days' scenarios. And as one of the most-persecuted religious groups in human history, Orthodox Christians certainly know what it means to prepare for hard times. But the grand testimony of the Church in her saints is that our greatest preparation for hard times is simply this: repentance, humility, and faith.[12]

The point seems to be the distance of god from everyday life; the greater the distance, the more people try to organize their lives methodically in anticipation of doomsday and their salvation. Weber's (1905) thesis on professional success as a proof of salvation emerges. According to him, rational preparation of every aspect of life was a way for people who embraced the Reformation to reduce the anxiety due to the rarity of signals of salvation. Food preparedness may be part this conception; not only is work a vocation (*Beruf*) but household activities must be prepared in detail, like a mission. Success in both cases is a symptom that the correct path to salvation has been taken.

The expectation is therefore that greater food preparedness is incorporated into the practices of people or countries with Protestant backgrounds. But demonstrating this, as at the time of Weber, is not easy. The phenomenon seems widespread throughout the European countries, although more so in the northern ones than those on the Mediterranean:

> The northern regions of the Nordic countries were especially characterised by what we call a 'storage economy.' Because of the harsh winters, where there was either snow on the ground or a lack of grass for grazing, it was necessary to ensure that food was preserved, such that it could last through the winter months. In many ways one can easily say that the dry, flat bread and *knekkebrød* crackers are a part of the storage economy. This kind of bread lasts a long time, 'If not from a child's birth to his wedding,' as Olaus Magnus claimed. We can also see the production of cheese and butter as a part of the storage economy. Even though both butter and cheese eventually had other functions, partly culinary and partly economic, the production of these two dairy products was one way of preserving milk. But it was meat and fish in particular that was preserved through drying, salting, smoking, curing, and fermenting, or most often a combination of the different methods. This is no longer something restricted to the aforementioned region. The storage economy was

---

12 https://blogs.ancientfaith.com/onbehalfofall/canned-food-mama-and-me/ (accessed 5 February 2016).

naturally a part of many regions of Europe, at least in those areas where there were mountains. The popularity Italian Parma ham and Spanish Serrano have reached today shows this well. The storage economy predominated in other parts of the Nordic region as well, even though it wasn't practiced in Denmark as much as in northern Scandinavia. In Denmark, smoking played an important role, as did salting. In fact, salted food dishes always pop up in old descriptions of food.

(Amilien 2012)

This long quotation provides a functional explanation of food storage: environmental conditions affect storage methods. Religious factors do not provide any inspiration. However, the quotation's reasoning introduces a new perspective on food storage: the combination of climatic conditions with the capacity to frame issues. Storing food – and in more general terms, processing it – is a way to define reality according to a cognitive register. For eschatological faiths, it is the means to identify practices of salvation; for other cultures, it may be a means to establish clear membership of a group. That was the explanation that Mary Douglas (1966) gave for the Jewish prohibition on eating pork. The pig is an animal with an unclear classification; it is a confusing entity unable to represent a distinct group membership. Purity is therefore a matter of cognitive clarity, or the capacity correctly to allocate objects, animals, and human beings to precise classes. Furthermore, such frames become practices of inclusion or exclusion of others, respectively as table-mates or non-invited persons.

Applying this way of thinking to food storage, processed food marks a sharp distinction from its natural origin: 'The rituals of cooking and table manners are frames that carefully prevent us from considering food in its natural origin and making us focus on its cultural attributes' (Coff 2006: 88). These attributes – of which storage techniques are an integral part – are further means to endorse a distance from others, who are then aliens, strangers, barbarians. The simpler and more intelligible the storage technique, the better it works as an identity marker. Anthropologists have interpreted storage as a crucial development in the passage of humankind from a natural state to a civilized one (see Chapter 1). Now we can add that food storage can also represent a line of discrimination with other social groups.

This culture-based interpretation – storage as distance from *wilderness* and wild people – is plausible, but it clashes with a recent robust trend which privileges organic and fresh food, seen as closer to a natural state. Historically, stored food represented a typical human way to keep a healthy distance from wicked nature. Instead, the modern ecological movement seeks to blur the nature/culture distinction, or to include the latter in the former. Several attempts have been made to elaborate less conflicting views of the nature–culture relationship – ecological modernization, shallow ecology, co-evolution (Pellizzoni 2015) – all of them appearing weak and provisional. Furthermore, some storage techniques – the most ancient ones – are seen as compatible with a harmonious, nature-driven life. We thus understand that the capacity of cultural symbols to define reality has changed through history.

Today, the primacy of culture over nature has been entirely overturned in the most radical ecological trends, or weakened in those groups pressing for less invasive or bio-imitating processes. All of them are linked to the movement for organic food, resistance to GMOs, and opposition to certain storage techniques like irradiation.

In conclusion, symbolization of food storage must be viewed in terms of the changing use of the nature/culture dichotomy. Today, the preservation of food can be recapitulated in four symbolic universes (see Figure 2.3): (a) an identifiable process able easily to mark membership of a social or religious group; (b) a way to conceive existence, emphasizing the idea of preparedness for an uncertain future, whether connected or unconnected to a religious event; (c) an emphasis on conviviality: stored valuable food is served for important celebrations, and (d) emphasis on bio-imitation: food storage makes it possible to establish a milder relationship with nature, for example by using its products according to the season.

The last meaning has already been considered: it is within the mainstream ecological movement, with a special emphasis on food storage practices that have a low impact on the environment in terms of energy and water use and pollution. The issue has already been considered in regard to food safety. Storage techniques can be distinguished according to the presence or absence of artificial additives.

The third meaning (conviviality) has not been mentioned before, but it is probably the most universal cultural value; it is exemplified by the Asian food reserve practices termed 'solidarity groups' (see Figure 2.4). Some foods are transformed and stored with great care because they will be consumed on special occasions. They are taken out of storage for important guests or to mark the main transitions of life (sacrament, marriage, death), as exemplified by bottles of vintage wine. But they can include canned fruits, liquors and various types of jam. They can be exotic foodstuffs, whose value lies in the preservation of a good taste despite spatial-temporal distance. The great value of spices in the Middle Ages – which spurred many explorations and some wars – was due to their remote provenance and their capacity to preserve special food for a long time (and also to their strong flavour).

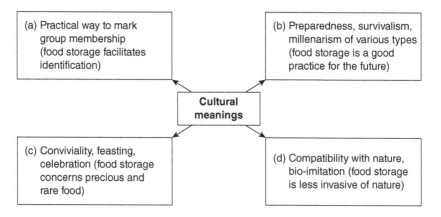

Figure 2.3 Main cultural meanings of food storage

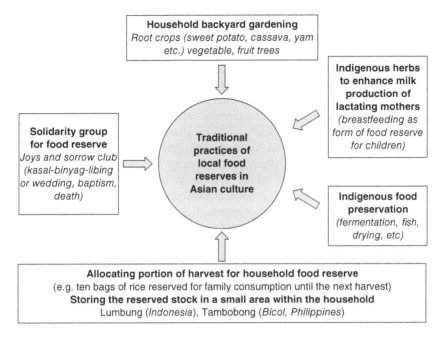

Figure 2.4 Traditional practices of local food reserves in Asian culture
Source: AsiaDHRRA (2011).

Although these various meanings are distinguished analytically, they are often mixed in empirical terms. This is the case with some survivalist movements, which incorporate relevant ecological ends (Mercalli 2012). Survivalism is a macrocosm of meanings and practices which do not comprise food storage alone. Survivalists prepare themselves for every aspect of life: the provision of energy, medicines, shelter, clothes, and of course food. A very divisive issue is the possession of weapons. Among the American preppers, having weapons is considered normal, perfectly in accordance with common sense (Lamy 2013). Among European preppers, this is considered a strange concept (Laudiero 2012).

The feature shared by all preppers is a *very negative analysis of the situation*; they expect the worst in diverse domains: climate change, bad government, social disorder and deteriorating international relations. Although the majority of preppers manifest or maintain a profoundly dark view of the world, they are determined to react rationally. Thus, preppers are a special category of persons; in a sense, they are optimistic in that they believe it is possible to deal with adverse events. The sociological suspicion is that they want to do this individually, without involving the public sphere, and without trying to address the causes of bad events at their origin.

This statement highlights a basic attitude of preppers: they are essentially self-confident, although they are ambivalent toward political institutions, which they see as static or, worse, morally corrupt. There are three dominant attitudes among preppers: an apocalyptic vision of the world, a positive evaluation of their own capacities to deal with adversities, and a sceptical opinion of public authorities. It would be wrong to say that they are fatalists, like one of Mary Douglas's grid-group types, because unlike people crushed by fate, their activism is in opposition to a perceived negative course of history. In conclusion, they seem to be 'ordinary people operating through cultural designs for anticipation and risk reduction' (Appadurai 2004: 64).

Analysis of the geography of preppers reveals more about their cultural biases. Between 23 February 2010 and 23 February 2014 – exactly four years – the website of the International Network of Preppers (http://www.international preppersnetwork.com) was visited 21,511 times by 12,536 visitors, according to the website counter. This leads to the observation that it is not a mass phenomenon, despite the fact that in the countries where preppers are more prevalent there is a long list of heavily frequented unofficial websites. In any case, the low numbers of visitors to this international website indicates that preparedness is a typical phenomenon in some countries. These countries are easily identified, thanks to the same website counter. The US, Canada, the UK and Germany accounted for 71 per cent of the total visits (see Figure 2.5).

To complete the evaluation of the data, it is necessary to weight the website visits by the population of each country. Thus, the ranking becomes:

1   New Zealand (8.9 visits per 100,000 inhabitants);
2   Canada (7.4 per 100,000);
3   Ireland (5.5 per 100,000);

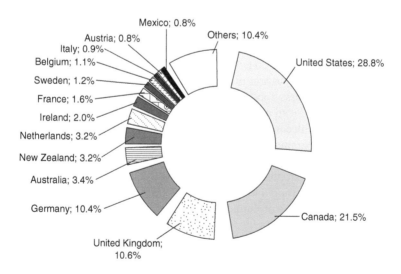

Figure 2.5 Percentage of International Network of Preppers website visits according to the country of the visitor

Source: http://s06.flagcounter.com/more/gv0/ (accessed 23 February 2014).

4   the Netherlands (2.3 per 100,000);
5   the UK (2.1 per 100,000);
6   Australia (1.8 per 100,000).

The USA, with its 3,563 visits, records a modest 1.1 per 100,000 inhabitants, but this figure is offset by the extraordinary variety of websites available in the country. In any case, all the top positions in the ranking are occupied by English-speaking countries, with the sole exception of the Netherlands. Other high-ranking European countries are Germany, Sweden, Denmark and Belgium.

Neo-Latin European countries have generally lower rankings in terms of visitors per 100,000. Lack of familiarity with English is probably the main explanation for this. Nevertheless, there are many signals that the preparedness mentality is rather alien to Mediterranean cultures. In Italy, large-scale food storage practices are seen as extravagant, or even as symptoms of neurosis. There is a prepper movement, but it adopts very moderate positions (see Laudiero 2012). As in France, Italian preppers try to distinguish themselves sharply from survivalists, seen as people practising an alternative lifestyle as if they were already in an emergency. Thus, the continuum presented in Figure 2.6 seems to be the best representation of this *symbolic universe* of people involved in emergency prevention actions.

Of course, this continuum is a simplification. Every prepper has his or her own history, sense-making system and set of daily actions. The purpose of the continuum is to highlight two aspects of preppers' lives: sociality and the sharing of practices. Preparedness is in the middle: a set of practices typically organized at household level, but with a social connotation, as the religious inspiration has shown (to be prepared for the others). In this sense, one extreme of the continuum is represented by civil protection volunteers: ordinary people periodically trained for emergencies, not to be confused with civil protection agency members, who are professionals permanently employed by the authorities to supervise and coordinate the volunteer corps before and during emergencies. At the other extreme of the continuum are survivalists: persons who feel themselves surrounded by a hostile environment, and evidently suffering from a reciprocal stereotype. For survivalists, the real point of differentiation is their evaluation of the world: the worse they perceive it as being, the more they insist

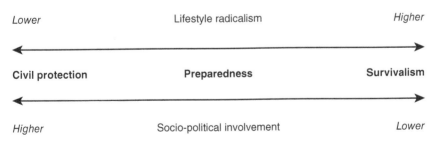

Figure 2.6 Locations of preparedness practices along the civil protection–survivalism continuum

on a separate lifestyle and sociality; the very extreme position is represented by the movie character Rambo – a very clever man, alone against the world.

We began with food storage, but we have finished with a broader field of inquiry. We have reached a point where food preparedness is part of a larger strategy of existence which includes cosmologies – the widest cognitive frames imaginable – and sociality, which strongly determines how everyday material life is structured. In regard to these aspects and the others considered in the chapter, it is possible to conduct a final appraisal according to the three aims of the entire book.

2.6 Rationing and rationalizing

The first question is whether storage is a category that is able to highlight new or neglected social phenomena. The answer is positive: the *food packaging industry* emerges as the actor able to reinforce the trend for individualization in food consumption. This chapter has highlighted the influence of distribution industry, including the industry of packaging and related services. It is a powerful sector with a great impact on the economy and the environment, and which has been able to lobby the institutions in a very subtle but effective way. Curiously, agrifood chain analysis tends to exclude the food packaging industry, while the real-life trend is that the combination of foods and packages is growing in terms of the variety of forms and sizes (Ahvenainen 2003). The storage and distribution industry is well equipped to support this trend with a wide range of devices, from the single-portion enveloper to the combined freezer-refrigerator. The industry facilitates food preservation for longer periods, but at the same time it provokes the enormous problem of disposal of all the materials and equipment created to process and deliver food. They are rarely recycled, but usually burnt, with all the polluting side-effects, or dumped in waste landfills that nobody wants close to them. The worst effects include pollution of the oceans (for instance, floating islands of plastic bags), and problems in poorer countries, where a miserable industry of recycling has developed (pollution havens). Hence food is stored in thousands of ways, but its waste is still an insurmountable problem.

The second question is whether storage been an ideal type with heuristic capacities. The answer in this case is also positive: the storage category has made it possible to emphasize the modularization of food processing – a process that will emerge for energy storage as well (see Section 4.6). The heuristic also serves to show the ambivalent effects of storage capacity on the agrifood chain, resulting in more centralization of command and more scattered production sites. There thus coexist property concentrated in a few hands – an oligopoly – and a network organization of production. Food storage replicates this dualism: decisions are made by a handful of persons, and it is practised in a myriad of places, including the remote areas of developing countries. However, this situation is insufficiently captured by the centre–periphery frameworks usually based on the Marxist idea of exploitation (Subramaniam and Bunka 2013).

The register of *ambivalences of food storage modularization* is more useful: on the one hand, new and small-scale preservation techniques are being discovered; on the other, their impact on human and environmental health is increasing and is a cause of major concerns. Safety control agencies are engaged in long authorization procedures in a sort of race with the packaging industry, which constantly provides new products to be tested. This process leads to bureaucratization rather than participation; it becomes such a specialized evaluation process that lay people are unable to follow, intervene and possibly decide. The storage techniques and materials involved in the food industry are considered less problematic by public opinion, and this further reduces participation. Public independent authorities, like the US Food and Drug Administration and the European Food Safety Authority, are designated to carry out the task. Unfortunately, only a restricted circle of specialists is involved with these agencies, and even then, there are suspicions they may be captured by private interests. Dialogue with civic organizations appears theoretically a good way forward, provided that third-sector problems of legitimacy (Fuchs, Kalfagianni and Havinga 2011), accountability (Scholte 2004) and financial independence (Wang 2006) are resolved.

The third question is whether storage is really the middle-range, intermediate perspective between deep and shallow views on environmental issues. Caution is necessary when addressing such a complex issue. Food storage, whatever actors' motivations, is a rationalization of consumption; it reduces the frequency of shopping; it allows the purchase of greater quantities, increasing the household's scale economies; it restrains impulsive purchasing, and it facilitates the development of a planned or reasoned organization of life. If storage is well organized, it also prevents food waste. The packaging industry and food storage techniques provide households with a wide range of products and services so that they can save money, time and environmental resources. Of course, this 'green economy' of food storage follows the individualization trend as well, by providing micro-food portions, whose storage demands a large amount of materials and energy. Industry manifests neutrality toward environmental issues; it is ready to serve the green cause, but it is also ready to provide every kind of item regardless of its impact on the environment. In any case, food storage has its own capacity of *rationing* and *rationalizing* (it is not by chance that these two words have the same root) to motivate, plan and discipline food distribution and consumption. This sort of enlarged rationality provides fertile terrain for fruitful encounters among producers, consumers and food authorities.

Food storage appears to be midway between the drastic reduction of consumption, as preached by the advocates of downshifting, and fast food as consumption without quality and discipline. Maintaining appropriate home reserves of food reduces waste and helps in planning consumption. In conclusion, food storage is not a humdrum activity reserved to zealous and dull housewives; on the contrary, as the flourishing of prepper circles shows, food storage provides an opportunity to develop imagination and sociality. This also applies to industry organization. The just-in-time fashion has contaminated the food industry by inducing the notion that warehouses are superfluous. Indeed, the preservation of food in large quantities is a major problem well known since antiquity; but it also fosters innovation – the equivalent of homemaker creativity.

3 Water storage
A multidimensional task

In this chapter, the storage issue is seen (1) as a historical movement against stagnant waters (marshland), (2) as a matter of flood or drought prevention (water security), (3) as a set of practices to conserve water in dwellings (harvesting), and (4) as a mechanism for backing multifunctional agriculture (pond networks).

3.1 The long march toward water channelling

In human history, there has been a general and progressive tendency to reclaim land from water in particular, although the term 'reclamation' includes the clearing of woods and the cultivation of stony soil. Curtis and Campopiano write:

> Land reclamation is the process by which people bring 'unused' or 'waste' land into 'productive' use. In the pre-industrial era, this meant the clearance of woodlands and bushes, the development of irrigation systems, or the drainage of wetlands, in order to create new land for cultivation and settlement.
> (Curtis and Campopiano 2014: 93)

Thus, land reclamation, like food storage and agriculture, appears to be a primordial activity of human beings, linked to their drive for settlement and sustenance. However, unlike food storage, the reclamation of land from water is a more shared activity: a social enterprise, involving a clan, an entire village or an authority.[1]

Water is seen, like wood and rocks, as a competitor for usable land. The frame in which to include this topic is broad and reaches far back in time, at least in countries of old civilization. In Italy, the Etruscans and Romans started extensive

---

1 There is a further political meaning of land reclamation: the claim for access to or use of land in the hands of absentee landlords or public bodies (see Moyo and Yeros 2005). In Italy, this was reflected in the movement for the occupation of land (Ginsborg 2003); in the UK, there is a specific policy: 'the Community Right to Reclaim Land [that] helps communities to improve their local area by giving them the right to ask that under-used or unused land owned by public bodies is brought back into beneficial use'; https://www.gov.uk/government/policies/giving-people-more-power-over-what-happens-in-their-neighbourhood/supporting-pages/community-right-to-reclaim-land (accessed 10 July 2015).

water management works. A well-known feature of the Po valley that still exists is the *centuriazione*: division of land into regular plots to be given to centurions as a reward for their service during war campaigns. The early Middle Ages is generally considered to be a period of re-swamping, which demonstrates that reclamation is not always progressive and irreversible. An increase in reclamation occurred in the late Middle Ages as a result of actions by monasteries and landlords. This long phase is also defined as pre-industrial, when reclamation proceeded sporadically. In peninsular Italy, the separation of Chiana swamp waters between the basins of the River Tevere, which flows through Rome, and the River Arno, which flows through Florence, lasted hundreds of years until the nineteenth century.

The industrial age did not arrive in every European country in the same period, but it sooner or later brought the use of steam-driven machines fuelled with coal and able to perform a great deal of work, and the era of mechanical reclamation began:

> Steam appeared in the Venetian countryside as early as the beginning of the 1830s, when the architect Giuseppe Jappelli convinced the baron Gaetano Testa to invest capital in reclaiming the Canale dei Cuori using his latest invention, a special aspiration suction pump: the *smergone*.
> (Consorzio di Bonifica Adige Euganeo 2013: 8, my translation)

Thanks to this innovation, the capacity to remove water from reclaimed land increased enormously. Such machines were able to drain land more quickly than the previous systems. However, it is correct to consider the improvement of reclamation not only as due to the introduction of a powerful source of energy, but also the combination of pumping devices and water channelling techniques. For example, in the nineteenth century a special type of dewatering pump using an aspiration turbine was introduced in the Venetian plain (Veronese 1925). This was an innovation imported from Holland. Previously, the mechanized drainage system consisted of blades mounted on a wheel rotated by animal power. This innovation was combined with the water flow management techniques known at that time.

Traditionally, there were two reclamation methods: drying (in Italian *essiccazione*) – simply moving water to lower places or into channels – and flooding or landfill (in Italian *colmata*) – diverting murky water into a basin where it was left to allow the solids to settle. The land level gradually rose as a consequence. But reclamation does not involve the application of a single mechanism. It is a package of techniques combined in a variety of ways – a *socio-technical system* similar to that described earlier for food storage. Mechanization and powerful energy sources – after steam, the advent of electricity led to rapid improvements in pumping capacity – were combined with the building of higher banks, because it was easier to pump water into the channels than to raise the land level with the landfill method. This had important consequences for the reclaimed landscape: it meant that land could remain dry even below sea level and in the absence of steep slopes towards major rivers and the sea. The capacities of these machines gave rise to a less ecological solutions for water drainage – ones that were insufficient when energy to drive the machines became expensive or heavy rains came. In other words, the combination of cheap energy provision and the invention of turbine pumps created a *technological path* whose consequences lasted longer than its usefulness.

Land reclamation is so important in places like northern Italy and the Po valley that it can be used to identify two epochal social processes: the beginning of industrialization and the development of social class relationships. In other words, reclamation entered the labour–capital dialectic. Industrialization in northern Italy began in two geographical locations: in the lower parts of mountain valleys, especially where falling water energy could be exploited, and around the towns situated at the foot of the Alps and Apennines; these were areas that had been colonized since Roman times, and it was easy to keep them drained because of the natural slope of the land toward the Adriatic Sea. Such areas were suitable for both settlement and agriculture: they were sufficiently high to maintain good water drainage, and sufficiently flat to benefit from spontaneous field irrigation. This is an important point of historical synthesis, highlighting the ambivalence of industrial and civil settlements toward water: it must be close, but not too close.

If we look at the definition of *water security*, the same ambiguity emerges. It is a definition that highlights the twofold attitude toward water: as a source of danger and a source of well-being. This ambiguity also explains why in the competition between water and land typical of reclamation, the majority of governments were in favour of the latter. Land is more solid, less capricious, and apparently more productive. For the same reason, there were long-term initiatives to drain marshes. These are intermediate between water and land, a mix of solid and liquid. Especially in the industrial age, their hybrid status has been seen as unacceptable because of the difficulty of calculating their revenues. This is even more evident in the case of the industrialization of agriculture: the systematic exploitation of land, first with machines, then with chemicals, requires a flat, dry and clear surface. Only when the ground has a certain consistency is it possible to make a correct calculation of its value and invest in it.

Reinforcing the bad economic image of marshes is their perception as sources of terrible diseases, typically malaria. In fact, it is possible to conduct a cognitive analysis showing that swamps are places that are difficult to define and accept. The fight against them is a sort of mission pursued by all Western countries, which illustrates how widespread the hostility to them is, and how deep their *historicity* lies. Historicity is a concept developed in terms of the landscape by Emilio Sereni (1961), referring to the overlap between long-term human–environment relationships in rural places. There are three main ingredients in the epic of land reclamation: (1) an enlightened authority figure, a strong character with the power and the capacity to conceive a grand reclamation plan, (2) a net gain of land and its distribution to promote the prosperity of the entire community, and (3) a fight against adverse natural forces, be they floods or diseases of marsh origin.

In Italy, the fascist regime greatly emphasized land reclamation. It used not only the instruments of rhetoric and mass communication, but also large-scale mobilization of people and means (Caprotti 2007: 229–230). The same had happened, for example, in Prussia two centuries earlier (Blackbourn 2007): a king with great ambitions, a wide area involved with the use of large-scale means, and the subjugation of natural forces through rational agriculture. Though with less emphasis on sovereignty, North America underwent the same process during the 1700s: 'settlers, commercial interests, and governments agreed that wetlands presented obstacles to development, and that wetlands should be eliminated and the land reclaimed for other purposes' (Dahl and Allord 1997). In Europe, these great reclamation projects brought a new balance to land ownership and labour relationships. Because the projects realized in the eighteenth century were very large in scale, great amounts of capital were necessary, and investors obviously wanted to become owners of the reclaimed land. This process, together with the intensification of labour on the new land, generated a new wave of accumulation and the formation of a huge mass of proletarians, the original nucleus of the workers' movement (Franklin 1969; Monti 1998).

This process is evident along the River Po, where large-scale reclamation took place during the nineteenth century with capital provided by large investors; the new land was then exploited using day labourers, called *braccianti* in Italian: people with no means but their arms (*braccia*), a denomination perfectly applicable to the Marxian figure of the proletariat (people whose only means are their offspring). In fact, a large part of the nascent workers' movement consisted of agricultural day labourers, who at the end of the nineteenth century violently clashed with landowners, mainly on land recently reclaimed from waters. Reclaimed land areas thus became flashpoints in terms of public order and attempts to control the socialist movement, places where the class conflict reached its highest levels. Testimony to this, at least in Italy, can be found in the frequency and severity of strikes in the countryside along the River Po. These were areas where the socialist party greatly increased its representation on local councils and, especially after the First World War, gained control of many municipalities.

For this reason, when a minimal level of democracy was restored in Italy after the Second World War, land reclamation schemes were usually coupled with land reform, in an attempt to reduce the usually uneven distribution of land resulting from the energy-intensive mechanical reclamation processes. The coupling of reclamation with land reform, at least in the River Po delta, worked to only a partial extent: it gave rise to a specific pattern of intensive or industrialized farming following a drive for land rationalization that began in the century of the Enlightenment (Carpanetto and Ricuperati 2008). However, it did not prevent a later new concentration of property in the hands of landlords – an outcome that occurred in many other locations as well:

> The cost[s] of carrying out land reform were often increased by the continued existence of implicit and explicit distortions which drove land prices above the capitalized value of agricultural profits and made it attractive for land reform beneficiaries to sell out to large farmers, thus contributing to reconcentration of holdings.
>
> (Deininger 1999: 653)

The last two centuries of land reclamation also affected economic development, at least in Italy. The areas reclaimed later thanks to the capital of landlords and the bourgeoisie were also those with the least growth in small- and medium-scale industry during the Italian economic boom of the 1950s and 1960s. The recently reclaimed lands were dominated by large estates, a new version of the old *latifundo* of southern Italy, the difference being that recent farms were run in a more managerial manner. The coupling of reclamation and the rationalization of agriculture produced in Gramscian terms a new *social formation*: a complex of production modes, state actions and cultural frames (Green 2013: 96). Newly reclaimed land represented a social formation centred on highly mechanized large farms concentrating on extensive crops like cereals, often combined with intensive livestock rearing in sheds, with the addition of scant interest in industrialization.

We are now able to frame reclamation within a broader process of agriculture modernization in which water became more specifically a means of production. It had to be channelled as much as possible so that it could be used for irrigation, removed from the fields as quickly as possible during the rainy season, and reintroduced at the same speed during the summer. The problem is that recently reclaimed land close to rivers and the sea is also land with the least gradient. In these areas, drainage and irrigation are more difficult: water tends to remain after abundant rains, and it is costly to irrigate the fields during the dry season. There are thus two kinds of land in the Po valley: land that is drained and irrigated spontaneously using only the slope of the terrain, and very flat land where water does not flow easily that is very costly to manage.

In land of this kind, the fight against water is harsher and constant. Stagnant water is the worst enemy to combat. The cultural bias against every form of stagnant water is evident and widespread among common people, too. According to the Gramscian theory, it stems from a manipulation of production forces interested in the above-mentioned extensive agriculture. Investors in highly mechanized farming have an interest in showing the uselessness or danger of steady waters: a sort of anti-stagnant water ideology (Giblett 1996). It was not an ideology against water itself, but against its immobility and its tendency to occupy fertile land. In positive terms, it was an ideal of water continuously flowing. It is not an overstatement to refer to *water flow* as an ideology because its encompassing and widely acknowledged nature. Agronomists and engineers have been the main proponents of land reclamation ideology in the industrial era, as the Benedictine monks were in the Middle Ages.

Thus we have a very broad picture of water and land management across several centuries. It comprises the idea of a *hegemony* of water flow against water stagnation. Hegemony, again a term from Gramsci (2014), stresses that authorities' expectations provide an impetus toward a new mode of production and a rationalization of knowledge that converge in land reclamation. This process has represented a long-term drive of development in all Western countries. Over the centuries, thousands of hectares were freed from water. Ponds of slack water remained small and residual. The only broad wetlands in Europe are river deltas, lands like those of the Rhone, Volga or Danube – a geography that demonstrates how pervasive the action against stagnant water has been. Hegemony over marshland can thus be considered quite successful, at least in the Western countries. Resistance to drainage activities did not exist, though some wetlands were preserved because landlords considered them useful for hunting. This is the history of royal game reserves transformed into nature parks in the twentieth century.

The history of the Horicon Marsh in Wisconsin is very informative:

> Horicon Marsh was dammed, flooded, and renamed Lake Horicon in 1846. At that time, it was the largest manmade lake in the world (about 4 miles wide by 14 miles long) [. . .]. Lake Horicon was used for commercial transportation and for commercial fishing. In 1869, the dam was removed and the land returned to marsh. In 1883, two sportsmen's clubs, which leased the marsh area, reported that 500,000 ducks hatched annually in the marsh. They also reported that 30,000 muskrats and mink were trapped in the southern half of the marsh. Huge flocks of geese also were reported [. . .]. In 1904, attempts were made to drain the marsh and sell the reclaimed land for truck farms. Lawsuits resulting from inadequate drainage halted the reclamation effort. In 1921, local conservationists began efforts to protect Horicon Marsh as a game refuge, and the State of Wisconsin created the Horicon Marsh Wildlife Refuge in July 1927. Later, to avoid legal confrontations with the local farmers, the State bought property and (or) water rights to the southern half of the refuge and the Federal Government purchased rights to the northern half. In 1990, Horicon Marsh was added to the sites recognized by the Convention on Wetlands of International Importance especially as Waterfowl Habitat.
>
> (Dahl and Allord 1997)

This history is interesting for three main reasons: (a) as in Europe, protected areas were originally game reserves for the upper classes; later, open-air activities diversified land use, but maintained an ambivalent attitude toward impracticable wetlands; (b) marshes were considered obstacles to rapid transport and high productivity; better alternatives were either land reclamation or *water replenishment* by means of dams; marshes therefore remained a hybrid terrain that nobody liked; and (c) even if protected by law, marshes were constantly under assault from small or large projects; the pressures and conflicts concerning marshland declined only with the public purchase of areas and usage rights.

The Horicon Marsh is the exception that proves the rule; abandonment of reclamation and the return of land to swamp status are rare. The first opposition to the hegemony of reclamation was raised by the ecological movement of the 1960s; as we know, it was initially concerned with air and water pollution and wildlife extinction: 'The first Earth Day (1970) led to the creation of the United States Environmental Protection Agency and the passage of the Clean Air, Clean Water, and Endangered Species Acts.'[2] The Clean Air Act did not greatly affect wetland

---

2 http://www.earthday.org/earth-day-history-movement (accessed 9 April 2015).

protection, because marshes were still seen as sources of *bad air*; on the contrary, the Endangered Species Act gave strong impetus to this protection because it was easy to demonstrate how rich in biodiversity wetlands are (see Chapter 5).

The symbolic date for marshland redemption is probably 1971, the year of the foundation of the Ramsar Convention, a treaty officially named The Convention on Wetlands of International Importance Especially as Waterfowl Habitat. The name reflects the original emphasis on the conservation and wise use of wetlands primarily as habitats for waterfowl (Farantouris 2009). Its special attention to birds was linked not only to their charismatic appeal (especially migratory species), but also to the older practice of hunting them.

Nevertheless, the break in reclamation hegemony has been established, with the main effect of ending most public subsidies for this practice. In the USA:

> the Highly Erodible Land Conservation and Wetland Conservation Compliance provisions (Swampbuster) were introduced in the 1985 Farm Bill [. . .]. The purpose of the provisions is to remove certain incentives to produce agricultural commodities on converted wetlands or highly erodible land.[3]

The conclusions are twofold. One is that areas with slow-flowing waters have been considered throughout history as among the worst situations for common people and for authorities with civilizing ambitions; reclamation is in fact not the suppression of water, but its channelling: keeping it flowing and establishing a clear water/land distinction. The entire philosophy of artificial river banking derives from this principle. The other conclusion is that the beneficial environmental qualities of marshes were discovered only few decades ago, after mediation with game reserves and special attention to birds. It was in any case a *cessation* of the ongoing process of reclamation, not a *re-conquest* of water on land. The already reclaimed land was definitively given over to agriculture and other urban uses. A return to unhealthy swamps appears theoretically and practically unsustainable. Indeed, examples of re-swamping are rare and considered only for 'new lands' – those in river deltas or along the coast.

The two conclusions outlined above recall the cultural bias against land/water mixing. It is not clear whether this is also an anthropological trait that induces people in every part of the world, and in every period, to channel water and thus separate it clearly from land. Detailed research on people living traditionally in deltas or fishing in marshes would probably reveal different scenarios (Saleh 2012). It has been demonstrated that in the history of Western countries, marsh economies based on fishing and the collection of fruits and herbs were widespread and kept separate from reclamation projects (Casagrande 1997). The question concerns the historical interpretation of nature productivity: in many places and at

---

3 USDA Natural Resources Conservation Service, 'Wetlands Conservation Provisions (Swampbuster)'; http://www.nrcs.usda.gov/wps/portal/nrcs/detail/national/programs/alphabetical/camr/?cid=stel prdb1043554 (accessed 9 April 2015).

certain times in history, reclaimed land has appeared to ensure better yields. This was certainly not only a matter of immediate profit. People combating marshland saw an opportunity to plan their activities more easily in a healthy environment. The hegemony of reclamation must thus be incorporated into an all-encompassing process of *rationalization* in the terms envisaged by Weber. Control of labour, control of nature and control of human bodies advanced together.

3.2 River banks, detention basins and floodable areas

The previous section ended with a pessimistic statement: residual wetlands are difficult to preserve; to think of increasing them for social and environmental reasons is foolish. The strong protection of property rights, the tight web of urbanization, and the unresolved negative attitudes toward combined land/water areas suggest that the return to marshland is impossible. But new situations may overcome the barriers to such projects.

Floods continue even in countries with good traditions of water security management. New extreme weather events are forecast, especially what the media call 'rain bombs'. In response to these possible events, reclamation and channelling have created a system of water management that is quite reliable, but also rigid (Hartmann 2011). This gives rise to a contradiction that is internal to the water management system: the sharp separation between solid (land) and fluid (water) increases predictability, in the sense that in stable external conditions it is easy to forecast land performance and water speed; this in turn makes it easier to establish the investment necessary to improve land and water works. But if the external conditions are not stable, such a water management system immediately becomes risky. If, for example, the quantity of rain is exceptionally high in a short period, the speed of expulsion of the water through firm, high-banked channels is insufficient. Water overflows the banks, or may breach them because of high pressure. The policy of cementing the internal sides of banks aims to counter this: to create a sharp distinction with land in order to allow farmers to cultivate even the tops of banks, and to increase the speed of water so that it leaves a specific area as soon as possible. Evidently, if every area adopts such acceleration measures, the water will very rapidly arrive at the final exits – beaches or coastal marshes – creating great pressure on those eco-systems. For example, the sea's capacity to absorb water from rivers can be limited by wind effects. In the upper Adriatic, this situation frequently occurs because of a south-easterly wind called the scirocco, combined with other factors such as high waters in the Venice lagoon. Another example is that the rapid arrival of fresh water can radically dilute the brackish waters of river mouths, altering local ecosystems and affecting the cultivation of clams and so on.

If stable climate and weather conditions are predicted, the water system can continue with the present strategies of sharp separation and water acceleration; but if more frequent extreme weather events are forecast even in temperate latitudes, the system is very risky. Other circumstances can contribute to increasing the turbulence of environments: for example, the massive presence of coypus, non-native aquatic rodents similar to beavers, in the ditches of northern Italy is considered a serious threat to watercourse integrity because they dig networks of tunnels which undermine the banks. Moreover, they have no natural enemies, so controlling their populations with artificial means alone is very difficult. Of course, climate conditions present the main difficulties for watershed management. The challenge, in fact, concerns the meaning and reliability we attribute to the expression *climate change*.[4] If it entails a gradual increase in sea level, internal water management systems can meet the challenge by enhancing pumping systems in lower-lying areas. But if climate change means an increase in extreme events, especially rain bombs, the current hydraulic security system is too rigid, and unable to absorb great quantities of water rapidly.

The problem not only concerns water, but the practicability of agriculture as well. If cultivation is organized with soil that is less able to absorb water, extreme rain events will severely damage the crops. If land is used solely for cultivation, leaving less and less space for riparian vegetation and ditches, extreme rain events will rapidly bring waters to the main channels, greatly threatening their capacity to withstand the pressure.

Of course, the rigidity of water security systems is variable. It depends not only on the management of flows, but also on the morphology of the territory and its urbanization. Nevertheless, there are evident patterns in the history of water security. The first has been presented above: it can be called *land separation and water acceleration*. According to this model, there are two main actions: (1) strengthening and raising river/channel banks, and (2) dredging river beds and securing the internal sides of banks through a range of methods: weirs, vegetation clearing, river deepening and culverts. The central feature is the bank; its solidity provides the main protection against flooding (Hartmann 2011).

A variation of this pattern is the creation of artificial containers to be filled in the event of floods. These are not new systems, because main rivers usually have strips of land alongside which are protected by banks. These are called *floodplains* (*golene* in Italian).[5] At frequent intervals along the River Po are

---

4 'Climate change' is a generic expression indicating 'the long-term fluctuations in temperature, precipitation, wind, and other aspects of the Earth's climate' (http://globecarboncycle.unh.edu/CarbonCycleGlossary.shtml, accessed 27 January 2016); but if we introduce the more connotative expression 'global warming', the arguments become more controversial (see Pettenger 2007).

5 'Floodplain' appears to be the most widely accepted translation of *golena*. A floodplain is defined as 'a strip of relatively smooth land bordering a stream and overflowed at a time of high water' (Baily 2005: 325). In the UK the official term used is: 'Flood storage area. A natural or manmade area basin that temporarily fills with water during periods of high river levels' (Environment Agency 2009: 1). However, a *golena* is sharply bordered – an area between external banks and the low-water level of the river in dry periods – while 'floodplain' conveys an idea of openness: 'land affected by flooding associated with waterways and *open* drainage systems'; http://www.melbournewater.com.au/Planning-and-building/Flood-and-planning-schemes/Pages/Planning-scheme-overlays.aspx (accessed 18 April 2015; emphasis added). In Germany: 'floodplains are defined as potentially submergible riparian land. They encompass land between levees or embankments of a river as well as flat areas behind the levees, which could be inundated by an extreme flood or crevasse of levees' (Hartmann 2011: 2).

two lines of banks of different heights: a master bank, and a lower bank closer to the river course. For most of their length, master banks are very large with a road on the top. Thus, especially in the central part of the River Po, *golene* are very large and able to absorb great quantities of water. Because of their extension and the transient nature of floods, *golene* have been gradually urbanized and cultivated. Some recreational activities have even been organized on these floodplains: for instance, summer open-air discotheques, restaurants, fairs and political party festivals, including the memorable *Unità* festival of the former Italian Communist Party.

Apart from their cultural functions, *golene* are essentially instruments for the containment of floods. Hence, permanent human activities on them can cause problems in emergencies. People and animals must be moved away; some buildings or structures may obstruct the flow of water. In any case, the management of *golene* is an important variation of the first pattern. It encapsulates a different approach to flood management: slowing down the speed of water so that it invades areas beyond the usual watercourse. Sometimes, ironically, artificial detention basin projects may be proposed for land reclaimed in the past from a river, thus returning it to its original function. In northern Italy, there are some detention basins especially for the tributaries of the River Po that descend from the Apennines. These are more torrential, and traverse less steep terrain. The need for emergency containment is more acute in flatter stretches where large masses of water may suddenly arrive and exceed the capacity of measures to deal with them. The critical location for such basins is therefore the intermediate stretch of the watercourse where it leaves the mountains.

The location of detention basins is based on many criteria, and often strictly hydro-morphological considerations are not the main ones. Their location depends firstly on the 'urbanization web', and secondly on the capacity of the authorities to negotiate solutions with the owners of land affected by the proposed detention basin scheme. Whatever the case may be, the change with respect to the former pattern is evident: water should not be confined as much as possible; rather, special mixed land/water areas should be created. We will see that the best solutions involve the transformation of such areas into biodiversity reserves or tourist attractions. The main problem arises when the urbanization is so dense that basins must be created among buildings and infrastructure. In these cases, the basin will be deeper and provided with vertical banks – resembling a parallelepiped – in order to retain more water. However, this requires the excavation of large holes with concrete walls.

This solution, as one might imagine, tends to encounter strong local opposition. However, the alternative – permanent evacuation of residents from designated flooding areas – is even more unpopular. Unlike the first pattern, in this case the conflict becomes clearly territorial: each residential area seeks to oppose construction of the basin by claiming that other locations are more suitable. In the land separation and water acceleration pattern, these geographical conflicts are progressively moved toward the mouth of the river, especially when they are less densely urbanized areas. But if even those areas are colonized with buildings and infrastructure, especially because of beach tourism, they will be less willing to accept diktats from upstream zones. They may insist that floodwater be contained upstream as well; the flood risk must be more equally divided along the different zones of the river, amounting to a claim for *geo-hydraulic justice* (Zwarteveen and Boelens 2014).

In conclusion, this second pattern of water security is much more politically sensitive than the strategy of speeding up the drainage of water, which appeared to be a matter almost exclusively delegated to experts. This raises an important question: to what extent do some environmental issues remain in the hands of experts (*de-politicization*, as described by Pellizzoni 2011b), and when and why do they instead become important political issues? Before answering, it should be said that there is a third pattern whose level of politicization is even higher. If the key term for the first pattern was 'high banks' and for the second 'detention basins', the third can be named 'controlled flooding'. This is based on the assumption that even well-organized and spacious containment structures are unable to cope with rain bombs.

Embankments, even if they contain extensive floodplains, do not offer the degree of security that is necessary in a period of climate change. Some areas beyond river banks must be allocated for possible flooding. Coming up with appropriate terms to describe this process is not easy: one could be *floodable areas* (Liao 2012) – not in sense that there is a probability of inundation, but that these are *areas programmatically designated to be flooded in emergencies*. In fact, a synonymous term is *planned flooding*, because gates are deliberately opened in order to inundate an area previously designated to receive floodwater. This pattern appears to be socially hazardous, at least in areas with a high density of human activity. In unpopulated or wild areas, the problem of finding places suitable to be flooded is easier to resolve, involving only decisions about the trade-offs among species and ecosystems.

In more populated areas, this solution is viable in *agro-forested areas*: in other words, in rural or quasi-natural areas where trees and vegetation can resist being partially submerged with water for a long period (Verhoeven and Setter 2010). Marshes would be the best examples, but as we saw in Section 3.1, internal marshes are rare and have been under attack for centuries. Thus, non-food tree plantations – dedicated to the production of firewood, timber, biofuel and cellulose – may be suitable for programmed flooding. Rice paddies can also be used to absorb exceptional amounts of water (United Nations Environment Programme 2005); the problem is that flooding is rarely compatible with the seasons of rice cultivation (Wassmann et al. 2004). Floodwater may remain in the fields much longer than expected and destroy the crop. The problem becomes political in the sense that (a) the cost of such solutions must be estimated (Massarutto and de Carli 2014), (b) the available fields must be determined, and (c) responsibility for payment of compensation to the farmers must be established, falling on either the government through ordinary taxation, or the residents in the protected areas through a special fee or an insurance premium (Morton and Olson 2013).

An interesting aspect of the controlled flooding pattern regards house design through the application of the old principle of *pile dwellings* or *palafitte housing*. Common sense indicates that such structures were erected for defence against animals and human enemies. Venice's origin – built on pile foundations – is interpreted in these classic terms. But their probable origin was as an intelligent solution to allow settlement of an area with a high variation in water levels. Rather than living somewhere with uncertain access to water, it was wiser to make a more radical choice and stay where water was permanently present, even if at varying levels.

Palafitte housing is widespread among the poorer populations along large tropical rivers.[6] The reason for this kind of settlement is partially linked to water security: monsoon areas experience wide variations in rainfall and river mouths are heavily affected, especially when they consist of deltas. Where access to higher land is difficult for lower-class people, they may respond by installing their houses in the residual urban commons, exactly marshland waters:

> The ports of Yucatan are also centers for local summer tourism, which creates an elevated, strong property demand, especially for ocean-front lots. This demand, coupled with land scarcity on the narrow coastal barrier island, leads to a relatively high cost for urban lands. As a result, lower socioeconomic groups, immigrants and established residents alike, are forced to fill in the neighboring wetlands in a disorderly manner with trash, rocks and sand to create areas for new housing construction.
> (Dickinson et al. 2006: 205)

Curiously, this type of housing has also been developed in very expensive tourist resorts, one variation being the houseboat. The latter – either fixed or mobile – is mainly a subcultural phenomenon; even in developing countries it is considered a practice, more or less ancient, linked to a special lifestyle. Artists, bohemians or celebrities have also tended to spend at least some time in such housing. Their lifestyles created a fashion, which has given rise to a rental market for houseboats. Some have even been transformed into luxury hotels. Moreover, the price for renting or using such houseboats is usually high because of the scarcity of moorings, at least in more attractive or affluent locations.

The function of houseboats that interests us here – water security – appears to be rare. It is mentioned in regard to the Dutch town of Maasbommel, which 'is pioneering floating houses, with flexible connections for fluids and electricity; these are not primarily intended for travel (and tourism), but rather to be safe

---

6 More limited cases include the waterhouses of Hamburg-Wilhelmsburg, Germany, along the River Elbe. This seems to be a case of gentrification, even if it complies with the demanding ecological standards of the Passive House movement (Venolia 2011). Quite different is the US case of Greenwich, Connecticut, where houses on stilts are built or adapted to counter higher insurance rates for buildings in risk areas (Linskey 2013).

70 *Water storage*

against flooding'. The town has attracted close attention from architects and urban planners because it seems to be a prototype of a settlement based on the idea of floating on water instead of being anchored to stable land. A permanent floating home may be a solution in lowland settlements if high water level variations are predicted. One problem is the numerous pipes required to supply such houses with services; they need at least a fixed basement on the ground. This opens up the prospect of designing off-grid houses.[7]

The third pattern has received some initial signals of approval from the European Union. On 23 October 2007, repeated floods induced the European Parliament and the Council of the European Union to issue Directive 2007/60/EC on the assessment and management of flood risks. The directive incorporates the concept of moving from an approach based on preventing the occurrence of flooding by strengthening structural interventions for flood defence such as banks, dams and artificial containers, to one based on the assessment and management of flood risks. The change requires the following:

a   A *flood risk map* must be drawn up, combining 'the probability of a flood event and of the potential adverse consequences for human health, the environment, cultural heritage and economic activity associated with a flood' – in other words, a territorial translation of flood probability and magnitude.
b   A *flexible strategy* focused on prevention, protection and preparedness must be adopted. These are activities framed by time. Prevention concerns the above-mentioned structural actions undertaken outside times of emergency. Protection occurs when an adverse event is imminent and threatens a specific area/population; it consists essentially of monitoring and temporary measures (for example, deployment of sandbags). Preparedness – a sort of tertiary prevention measure (see Chapter 2) – regards the installation of devices that are useful after the disastrous event has occurred. Assistance from civil defence corps is fundamental for protection and preparedness.
c   All natural or artificial channels for reducing water pressure must be developed: 'With a view to giving rivers more space, they [plans] should consider where possible the maintenance and/or restoration of floodplains' and take into account 'relevant aspects such as costs and benefits, flood extent and flood conveyance routes and areas which have the potential to retain flood water [. . .] as well as the controlled flooding of certain areas in the case of a flood event'.[8]

---

7 Studies have been conducted in Holland on buildings that are autonomous in terms of energy provision and dealing with slurry waste (Van Vliet, Shove and Chappells 2012); these solutions have already been trialled in ships, trains and caravans.
8 All quotations in this list are from 'Directive 2007/60/EC of the European Parliament and of the Council of 23 October 2007 on the assessment and management of flood risks', *Official Journal of the European Union*, 6 November 2007, L 288/27.

A picture of flood management emerges that is more composite than the accelerated drainage and container construction patterns. Those interventions are not discarded; rather, they are incorporated into a broader strategy of flood impact evaluation. A significant passage in the directive concerns a cost–benefit analysis, indicating that some prevention measures may be too costly in terms of the damage they would cause, and there may be systematic undervaluation of the huge costs of post-flood repairs.[9]

Moreover, the directive suggests that floods are inevitable: 'Despite many efforts to protect against floods, it has proven impossible to eradicate them completely. For this reason attention in Europe has shifted in the past decades from protection against floods to managing flood risks' (Mostert and Junier 2009: 4,962). The directive is therefore a bridge between the detention basins pattern and controlled flooding, in which a new and uncertain space is created. This space is outlined in *flood hazard maps* [that] should cover areas that may be affected by floods with a low probability (extreme event), floods with a medium probability (return period ≥ 100 years) and, where appropriate, floods with a high probability ($\sim HQ_{10}$)' and *flood risk maps* that are 'qualitative risk maps which should show the number of potentially affected inhabitants, the types of economic activity, protected areas affected, and information on possible pollution source' (de Moel, van Alphen and Aerts 2009: 294; emphasis added).

According to this elementary classification of flood occurrences, territories can be organized in four main ways: (a) establishing limits on construction and infrastructure, (b) issuing insurance premiums to cover damage to people and things, (c) educating people how to react in the case of flood, and (d) creating a monitoring and communication system. These strategies can be part of a flood risk plan whose aim is to manage a virtual and physical space of uncertainty between everyday human activities and water flow variability, providing an interactive system involving residents, the authorities and water management experts. The most interesting aspect of such *risk management* is the permanent learning system. Because forecasting is very imprecise, a cyclical system of training must be maintained.

Permanent learning is not the only system for reducing uncertainty. Risk can be monetized by providing compensation for all the terrain included in a certain level of flood likelihood. This brings us to the central point: the so-called third pattern is ideally an open, explicit, compensated agreement between water authorities and the owners of land affected by the possibility of floods. The former assume responsibility for deciding when and where to open floodgates; the latter accept being inundated, with the periods, timing and duration of water coverage left undetermined.

---

9 In some areas of the River Po delta, the energy costs of pumping water off land are so high that it has led to speculation whether it might be more convenient to let those areas, today devoted to agriculture, return to marshland (see Galvani 2010: 187; for details of other locations, see Morton and Olson 2013, and Bruzzone 2012).

72 Water storage

Table 3.1 Diagram of the three patterns of water security according to social, political and economic aspects

Analysis criteria	Increasing the water speed	Water basin construction	Controlled flooding areas
Costs	High for ordinary bank maintenance and water pumps	Very high for detention basin construction	Depend on the amount of compensation
Conflicts	Low because it is a landowners/agencies win–win game	Limited with land owners for compensation assessment	Very high because of the indeterminacy of land occupancy
Efficacy	High when weather and water flow are stable	Higher than increasing the water speed, but doubtful for experts	Very high, even if hypothetical because of the lack of cases
Territorial equity	Low for lowlands where river beds are elevated	Low for middle stretches of rivers where detention basins are constructed	Low for controlled flood areas in the case of poor compensation

Some aspects of water security patterns have already been discussed; others warrant brief explanations. Conflict with local people is important, and varies between the three patterns (see Table 3.1). For the first, 'increasing the water speed', the assessment of low levels of conflict is due to the fact that accelerating the flow of water provides gains for every landowner: for farmers who can dry their fields very rapidly in order to work them easily, and for those on urban estates because they gain ground along the rivers. What has happened in towns is evident: the beds of rivers and channels were restricted with artificial walls, if not covered or pipelined. The lad areas thus acquired were used for urbanization. Not only did private developers benefited, but so did public authorities, which could allocate the gained spaces to public functions (parks, cycle paths, car parks and so on) or impose land taxes on private uses.

The conflict is of medium intensity in the second pattern, 'water basin construction', because the problem lies in removing land from its usual function in order to create the detention basin. It is true that detention basins can be kept green with terrain banks and have the possibility to be internally cultivated, but inundation uncertainty for landowners and farmers must be compensated. Conflict arises either because proprietors will not accept such uncertainty or because they refuse monetary compensation (or insurance cover) they consider inadequate for the damage they will suffer. Thus the conflict is more circumscribed in terms of the actors and areas involved. In the case of public land, the problem is reduced to which services to privilege within the detention basin. Other compensation sources can come from ecosystem services paid for by urban dwellers (Massarutto and de Carli 2014).

The possibility of conflict is much greater for the third pattern, 'controlled flooding areas', because controlled floods in open fields are more unpredictable in terms of frequency, magnitude and extent. Measures of ground altitude and the accurate location of floodgates can facilitate the prediction of inundatable zones, but uncertainty remains high because of the extent and openness of the land affected. Conflict concerns compensation not only for the risk, but also the psychological distress of being under threat of flooding. Of course, people living in such areas must be prepared for the possibility of floods, with access to both material facilities (tax breaks, monetary compensation, equipment and so on) and immaterial ones (access to learning procedures, public recognition and so on). In any case, the risk of conflict is very high – so high that this third pattern is quite rare.

Before considering some (rare) cases, another social aspect of these patterns warrants comment: *territorial equity*. Space inequality seems to be a permanent feature of each pattern, although it takes different forms. In fact, the maximum inequality for the first pattern concerns downstream areas near river mouths. These suffer from all the flooding risks of high-speed waters, so owners must protect their land with large embankments, both from the river and the sea. Gradually higher embankments may be installed nearing the river mouth, but with a higher risk of flooding because the bottom of the river is above ground level.

High inequality is also recorded evident in the second pattern, but in this case the most severely penalized area is the middle zone of the river, where detention basins must be located. Upstream there is usually no room for such structures, and downstream they are useless, so they must be located in the middle river course, usually upstream from a large urban centre or an important industrial area. The choice of location is restricted to a few areas, and is even more reduced for zones of intense industrial and urban sprawl. This is connected with the above-mentioned possibility of NIMBY (not in my back yard) localized conflicts.

Inequality is most insidious for the third pattern, because two social factors are at play: the political strength or self-esteem of a rural community under consideration for programmatic inundation, and the persuasion capacity of the public agencies required to demonstrate the utility and appropriateness of the designated floodable area. At stake is a demanding pact between town and countryside for the provision of a vital flood prevention service: urban areas have much more to lose in the event of extensive flooding, and should compensate the open or agriculture areas that agree to be inundated in their stead. Territorial justice and injustice are linked to a long-term agreement among areas with different political strengths and stakes.

In that sense, the situations are very different. There follow two illustrative cases: the first concerns the global South; the second a developed country (the Fargo Project in the USA).

> The pilot project 'Sustainable management of transboundary floodable forests in the Amazon Basin' is directed by Ms. Patricia Chaves de Oliveira, PhD in Agrarian Sciences, who works with a team of three Peruvian and three Brazilian consultants in the floodable forests of the Peruvian Amazon. These forests are subject to regular flooding from the Amazon

River in both Peru and Brazil. 'In Peru, floodable areas or *tahuampas* correspond to floodable areas or floodplains. This is the case of the region of Loreto and its capital Iquitos, which includes the provinces of Maynas, Loreto-Nauta, Ramón Castilla and Requena. Particularly noteworthy is the Pacaya Samiria National Reserve (home to one of the largest floodplains in Loreto) [. . .]. The project expert is particularly interested in ethnic knowledge, the way that the local residents face these regular floods. She explains ethnodevelopment as the knowledge, memory and culture of ethnic groups, used to seek and create adequate and sustainable solutions. Why is it important to manage the floodable forests of the Peruvian Amazon sustainably? Ms. Chaves de Oliveira points out that the biodiversity found in Peruvian flooded ecosystems is extremely varied, which offers the possibility of applying technology to use and manage it, generating income for tahuampa residents. This will make it possible to create new green economy scenarios in the Amazon.

(GEF Amazon Project 2013)

This case is interesting for many reasons: floods are considered normal; they can be seen as sources of income for local people thanks to the development of special flood forest products. The project derives from the encounter between expert knowledge from traditional sources (universities, UN agencies) and ethnic knowledge. Only economic activities are mentioned, while no information is given on residential ways to deal with flooding. According to images on the project website, we can envisage some sort of palafitte style, which appears very suitable in the context of the Amazon flood forest.

The Amazon basin is an extreme context, and it concerns only a few people involved through voluntary schemes. On the other hand, programmed floods can result in milder impositions on many people. This is probably the case with the above-mentioned rice paddy inundation to prevent the flooding of populated areas in Asia. It seems from the available evidence that in some countries with low democratic profiles and pressing socio-hydraulic problems, the order to flood rice fields is peremptory, and only gaps in the line of command from centre to periphery can give relative autonomy to rice cultivators (see Manuta et al. 2006). Moreover, farmers frequently suffer from the delay – or worse, the absence – of reimbursements for the damage caused to their fields by forced flooding. The loose control of centre over periphery is compensated with less recognition to farmers of damage caused by the flooding of their rice fields.

The controlled flood pattern then comprises a political problem: who has the right to decide the institution of the scheme, and which areas must be flooded in the case of peak discharges and storm surges that coincide? According to Meyer and Hermans (2009), it is politically noteworthy that: (a) the controlled flood pattern is not only a matter for rural areas, but should also involve highly urbanized areas; (b) the problem of participation by non-experts in the choice between different patterns is not contemplated. The case of Glasgow's drainage area plan illustrates both issues (Jones and Macdonald 2007).

Even the new adaptive pattern is usually an exercise by planners, who formulate hypotheses on controlled floods or 'living with waters'[10] without the participation of people. The deliberate inundation of fields or urban areas requires either wide consensus among residents or strong measures to adapt buildings and properties. Both conditions are difficult to achieve without a massive injection of public funds into the adaptation of buildings, for example with waterproofing techniques. Most of the literature on controlled flooding consists of top-down maps conceived in a very refined manner, for example using Google Image Search without the involvement of local people, who will probably rise up against the plan when they know they are zoned for possible deliberate flooding.

Two ways to escape from this expertise-driven strategy are to consider agreements with residents and landowners on programmed floods and/or consider mobilization for preparedness – a practice mentioned earlier with regard to the EU flood prevention directive. Again, the existence of plans and projects rather than effective agreements among stakeholders on controlled flooding is notable. Of interest is the second case, in a developed country, that of Fargo in the USA:

> As of April 24th 2013, for the fourth year since 2009, young and old in Fargo, North Dakota, were filling sandbags to hold back melting snow floodwaters of the Red River. Since 2008, Fargo and nearby Moorhead, Minnesota, combined, have spent an estimated $195 million annually for damages in those flood years. To end this crisis, Fargo's mayor rallied public support for replacing this reactive adaptation-by-sandbags with a pre-emptive, engineered adaptation to flood control. But local farmers and residents in surrounding communities are objecting to the engineered solution. The mayor proposes construction of a 36-mile long Corps of Engineers' channel, 100 to 300 feet wide and up to 29 feet deep, to divert Red River floodwaters. The channel would leave the river at a point north of the city, loop around to the west of Fargo, and then re-join the river south of the city. Such a channel would take 8 to 10 years to build at an estimated cost of $1.4 to $2 billion, mostly from federal money [. . .]. The residents of adjacent towns and farmlands, who would be disrupted by the channel, are arguing for a combination of wetlands and 'waffle' within the existing Red River floodplain. In this strategy, the river's original shoreline wetlands would be restored to absorb and slow floodwaters. The 'waffle' concept [. . .] would consist of low berms

---

10 'Historically, cities seeking to prevent flooding have built walls and levees to keep water out. Repeated flooding and levee failures along the Mississippi River, however, have led to increased focus on flood "resilience" (recovering quickly and relatively inexpensively from flooding) over maximum "resistance" (keeping water out). Seattle, WA and Charleston, SC, for example, are developing "floodable zones" that preserve the city's access to its waterfront while minimizing damage when periodic flooding occurs. This concept of "living with water" is an option to consider for Boston as well' (Douglas et al. 2013: 6).

built around fields adjoining the river to trap flood water in the fields, letting surface water percolate into the groundwater and siphoning off flooding. The cost of this approach is an estimated $160 million, plus annual payments to farmers, and it could be built incrementally beginning in a year or two.

(Steinhour 2013: 1)

This case has all the ingredients of our reasoning on patterns of flood adaptation. Furthermore, local people are mentioned not only because of their opposition, but also with regard to a positive (and cheaper) solution. Unfortunately, the case is a very recent one, and it is not yet known whether the dilemma has been resolved. Thus, good proposals, fine-tuned projects and democratically agreed plans representing the third pattern are found throughout the world. But they are all still at the potential level. There is probably an intrinsic reason for this.

Controlled flood areas are by their nature 'fuzzy', with unclear external borders. Because of diverse territorial morphologies, it is not easy to forecast precisely where and how the water will flood the designated areas. The same uncertainty surrounds the consequences of inundation. In the case of forest, moorland or wetland, the prediction can be easy; but for cropland, the results may be very variable: water can fertilize the ground, like Nile silt, or it can transport materials so polluted that they threaten the health of subsequent land users. In turn, all these uncertainties increase the difficulty of establishing compensation for landowners.

A more extreme solution in terms of adaptation is *self-preparedness*, resembling that described for food in Chapter 2. Because it is so difficult to establish the precise floodable area and floods are inevitable, the best solution is *household preparedness* – a type of primary prevention. Other than temporary barriers made of sandbags, there are more structural measures, such as waterproofing or elevating the house, or modifying the ground floor to cope with inundation. Floating housing – another measure – has already been discussed. Along the River Po, houses on *golene* were traditionally organized so that they could receive water which submerged the entire basement or cellar.

The problem with household-level preparedness is the cost of adaptation. In a country like Italy, where people have invested a large amount of their wealth in their homes and furniture, the changes caused by waterproofing measures or sporadic water penetration are very demanding. Of course, as we have seen for food, the cultural background matters. Lack of preparedness is evident in Italy in regard not only to flood risk, but also other sources of danger like earthquakes and car accidents. House adaptation costs could be subsidized by the public authorities because such interventions are certainly less expensive than traditional defence measures, as the Fargo case demonstrates. However, authorities and experts do not always prefer the most convenient solution. A large budget for water management means a greater capacity to distribute money to companies, professionals and clients. In conclusion, implementation of the third pattern of water security is impeded not only by its intrinsic uncertainty, but also by traditional socio-technical systems not always inspired by rationality (and honesty) alone.

3.3 The art of water conservation and harvesting

The capacity to store water in times of excess has proved crucial in terms of security. Another field where such an ability is important is water conservation. The general topic is well known: around the world, a great amount of fresh water is wasted, for two reasons: shortcomings in systems of provision, and overconsumption by affluent people. Of course, the level of wastage depends greatly on the availability of fresh water. In some semi-arid areas, its scarcity is the cause of disease, while in temperate latitude areas, rich with mountains acting as sponges, water is so abundant that 'land floats on a groundwater mattress' (Distretto Idrografico delle Alpi Orientali 2004). This is the case of many areas of the Po valley, referred to a number of times earlier in this chapter.

Water storage processes are very similar to those of food conservation; both concern vital resources badly distributed among humans. The difference is that water is more fluid than food, so it is more difficult to restrict its availability. It is closer to a *common good* (rival consumption and difficulty of exclusion), which requires a special form of control to prevent free riding. At the same time, modern water distribution by pipes originating from a centralized source creates the conditions for a natural monopoly: the two conditions – few sources, many branches – require public intervention in water distribution. In fact, throughout the world, with some local or country variations, potable water provision systems are under the control of an authority.

Another difference compared to food is that water is very difficult to transport over long distances. It *is* possible to transport water, but it is convenient only for small quantities of high-quality bottled mineral water. The issue of distance is crucial in highlighting the importance of storage. The Romans built long aqueducts with spectacular arches, still visible in Segovia or Istanbul. The problem was to how to convey safe water regularly to large human agglomerations. Storage was the complementary solution to water provision: 'the combination of cisterns with the impressive water conveyance constructions' (Mays, Antoniou and Angelakis 2013: 1,917). In Istanbul, it is possible to admire in the city centre the Basilica Cistern, a large underground reservoir built during the period of Justinian and able to contain 80,000 cubic metres of water. The storage of water was usually complementary to mains supplies, and exceptionally subsidiary of aqueducts. In fact, in Istanbul as in other places with periodic shortages of water, the underground cistern was a means to store a vital resource in case of siege.

For medieval towns built on the tops of hills for defensive reasons, the provision of water in large containers was crucial. Water descends quickly, especially in karstic rocks. Furthermore, in such situations it is very difficult to build bridge aqueducts in the Roman style because of the towns' elevation. Thus, either very deep wells must be dug, or water storage is the only way to provide water. The central Italian town of Orvieto probably provides one of the best examples of this situation: there are many small water storage caves under the town, and St Patrick's Well, so large that water can be brought up from it with wagons pulled by mules along two overlapping spiral paths.

Until a few years ago, in most cities of the Mediterranean the house cistern was the normal solution to seasonal water shortages (Mays, Antoniou and Angelakis 2013). It provided a way to be independent and gain prestige in the community. Richer people had more beautiful and more secure houses. Genoa, an important harbour city in northwest Italy, was traditionally very affluent. It had short rivers upstream, and was surrounded by low mountains unable to provide abundant fresh water throughout the year. Thus the city had to obtain water from other valleys more than 40 kilometres away (Guastoni 2004: 24) and to build numerous cisterns. The case of Genoa shows the coexistence in history of two provision systems. There was a private system for richer houses equipped with cisterns; such households had the right to divert some of the aqueduct water to their private domain in order to have regular stocks of water. The other system consisted of fountains and tanks providing free water to residents – what is technically termed a *public good*. Because the second system had no upstream reservoirs, it was vulnerable to flow variation in water sources. In other words, it was free of charge, but it was not secure.

This issue is why some residents of Genoa had the right to divert and store water privately. Probably of noble rank, they possessed land and exploitation rights on a variety of goods; they could consequently claim an exclusive right to water as well. Then, water storage measures went beyond simple ecological availability, giving rise to usage rules and the control of water by an authority. Rights over water and technologies for its collection and transport also became important because they eliminated the need to store water in places subject to periodic shortages. At the end of the nineteenth century, a technological revolution came about through three developments:

- the construction of underground cast iron water mains inside town and city walls;
- the determination to convey fresh water from very distant places with long pipelines;
- the capacity to purify large quantities of water to make it potable.

The combination of these advances confirmed the success of the constant distribution of water by networks or mains despite seasonal trends. The spread of this system produced two important innovations in the relationship between water and its users. One was the supply of water to almost all private dwellings, so that it was no longer a privilege of noble or affluent people. The other was the possibility to measure consumption, so that households paid for the exact amount of water taken from the mains.

This has been the evolution of potable water provision in the most developed countries, where water is abundant or reasonably close at hand, or the local society has been able to construct systems to convey it over long distances. US towns in particular may be served by aqueducts hundreds of kilometres in length. In these circumstances, the storage of water becomes pointless. Modern technological systems deliver water to every consumption unit – houses, public services, workplaces – in great abundance and at relatively low cost. The storage of water is thus polarized between large artificial upstream reservoirs and small bottles for daily consumption. In the middle, the network does most of the work: the distribution of potable water to thousands of users. The grid itself becomes a storage structure.

In poor countries with predominantly barren land, water is another matter; because there are no mains, the supply depends on wells and their accessibility. The mains work is done by people's feet as they carry water from a spring to their homes on foot or with poor means of transport. Furthermore, a well can be easily controlled through rights over the piece of land where it is located, and through possession of the devices (pumps, drills and so on) necessary to provide access to it. In other words, for legal or technical reasons a well is an easily excludable good. Provision from lakes or ponds may be less excludable, especially when they are very large and under different administrations. This is dramatically the case with Lake Chad, where easier accessibility is threatening its disappearance. The *tragedy of the commons* is ever-present – a risk which raises the question of what institution is most suitable for ensuring equal and sustainable water distribution.

But there is a third, apparently better, situation: that of the towns in poor countries traversed by large rivers. In these cases, the basic good is available, but there is a diversified supply system consisting of mains, public fountains and water tankers. The coexistence of different provision systems creates two problems: one concerns the quality of water because of contamination with sewerage and other sources of disease; the other is social inequality, in that poor people subsidize rich ones.[11] In this intermediate situation, accumulation of water is demonstrably working. The scale of water accumulation becomes a crucial criterion:

- accumulation using containers or cisterns at household level;
- cisterns for apartment blocks and terraced houses;
- communal water kiosks located in neighbourhoods;
- local management of fountains or wells integrated with a water tanker service;
- large reservoirs connected to the local network;
- dams upstream of the conurbation.

---

11 'The unconnected poor, black, or indigenous population has subsidized the public system by paying taxes or purchasing water in tankers at much higher prices, while the main beneficiaries have been the more affluent, white population who are connected to the public system' (Spronk 2010: 163). The picture is probably more composite than a rich/poor dichotomy: intermediate organizations, sometimes of criminal origin, speculate on water distribution by tankers, penalizing not only poor people, but the entire grid supply system. Southern Italy has long suffered from a similar situation (Giglioli and Swyngedouw 2008). In India, water cisterns and dikes have been controlled through a complex relationship among castes, intermediate groups and Untouchables (Martínez Alier 2009: 193ff.; see also Ranganathan 2014). In the Middle East the situation is better: most buildings have a tank on the roof and are connected to the grid. However, the problem is the political use of scarce water sources (Allan 2002).

80 *Water storage*

The intermediate situation therefore relies on water accumulators of many sizes; they represent an interesting and flexible integration of water storage systems and water pipelines or grids. In conclusion, three types of water provision can be identified:

1   no-grid provision systems, in which water storage is small-scale and practised at family level, especially by women and children;
2   partially covered grid systems, in which storage varies in terms of territorial scale and quality;
3   totally covered grid systems, in which storage is used only in large buildings (for example, hotels) for reasons of security (to meet peak demand and in case of fire).

According to this threefold typology, water accumulation is crucial only for the intermediate water provision type. In poor countries, the means to maintain large quantities of water in houses is lacking; thus, family members have to make daily or weekly trips to water sources. In rich countries, storage at single-building level is unnecessary because the grid supply is abundant and constant. In this case, water conservation is seen as a potential cause of various diseases. In the best situations, where water pressure in the mains is high and constant, even domestic water heaters may lack storage tanks: instant heaters, usually fuelled by gas, can immediately provide the required amount of warm or hot water.

Thus, accumulation is a crucial factor only in cases where water is relatively abundant, but the provision system is limited by various factors (lack of money to build the mains grid or to connect to it, poor quality of the water pumped into the grid, irregular supply through the day or the week and so on). One example of this is the description of the water supply in a large town in northeast Brazil:

> Public services provide almost all the houses with water supply, energy, sanitation, and regular garbage collection three times a week, which are regular and rarely interrupted. In some houses, old tanks for storing water from the rain can be found, reminiscent of a time when potable water supply did not exist in that region, or services were precarious. [. . .] A significantly high percentage of middle class homes has a private water supply system (well and/or water tank). The current method of storing water is the use of sealed or netted roof tanks, located in the ceiling under the roof of the house, that receive an uninterrupted potable water supply from the company responsible for the public water supply in Fortaleza.
>
> (Caprara et al. 2009: 129)

The situation is rather different in the poorer blocks of the city:

> Although the public system supplies water to over 80% of the dwellings in the under-privileged blocks, the people there face daily water supply problems. This situation is aggravated by the fact that some inhabitants have no plumbing in their homes, whether due to the absence of public services in the

Water storage 81

dwelling or because not all families can afford it. [. . .] Thus, the inhabitants buy water from other sources, such as buying in bulk from horse-drawn carts; from motorised tanks; or from people who walk around the streets with large cans of water. This irregularity (or inexistence, in some cases) of water supply from the public sphere leads the population to store water in various containers such as water tanks, cisterns, barrels, drums, bowls, pots, water filters and others. [. . .] It is a poor population that cannot afford to have their own roof tank and in order to solve the water shortage problem, has acquired the habit of waking up daily at dawn to open taps and fill up buckets and drums.

(Caprara et al. 2009: 131)

The two quotations are taken from an article on Dengue fever vectored by a mosquito that breeds in stagnant water. The problem for poor people is that they may have a variety of water containers, neither sealed nor regularly filled, which provide conditions that facilitate the spread of mosquito-borne and other diseases. It is clear that water storage is very useful as long as it is organized appropriately: for example, using large sealed tanks installed under the roof and connected to the grid or to a deep well.

The Fortaleza situation corresponds to a typical social justice frame: the water supply must be guaranteed to all people everywhere; such a system prevents the spread of certain human diseases. This is the hygiene practice and the ideology dominant in the European countries since the end of the nineteenth century (Lewis 1952, quoted in Jones and Macdonald 2007: 536). But the picture is complicated by two factors. One is the legitimate aspiration of people in poor and emerging countries to have their homes connected to the grid; the other is the overconsumption induced by mains provision even in places less rich in water.

Aspiration is a relative concept; the quantity of tap water available varies according to individual preferences and reference groups (Escalas and Bettman 2003). In the most advanced countries, a minimal quantity of water per day is usually established and supplied. But it is a threshold enforced only by public services for people who are unable to pay the fee. In fact, most people consume a much larger amount of water. This phenomenon of increased consumption due to ease of accessibility thanks to improved technology is known in one sense as the *rebound effect* (Brookes 1990) and in another *rising expectations* (Gale 1987). It happens very frequently, even in places where there is restricted availability of a particular good. Hence, a water grid is built, but simultaneous consumption by many users leads to under-provision, if not discontinuation of the service. The standards of water public services have to find a balance between the aspirations of water users and the availability of water in the environment.

Faced with rising expectations, there is a demand to increase the system's capacity by pumping more water into the mains. However, this entails increasing costs, for three main reasons: (1) water must be sourced from more distant areas that may deny permission or demand monetary compensation; (2) more water in the grid leads to an increase in the need for sanitation treatment, which is very costly, and (3) energy costs for pumping and grid maintenance increase: 'The water sector consumes 3%

of the electricity generated in the U.S. and electricity accounts for approximately one-third of utilities' operating costs' (Kloss 2008: 8). Usually, public potable water services are unable to cover these costs, so they seek external sources of finance. Private companies can furnish new capital; but in exchange for it, they want a share in the water supply agency's ownership and demand a guaranteed profit rate. This is the reason for the tendency to privatize civil water services in many countries – a measure fiercely opposed by social movements (Molle, Mollinga and Meinzen-Dick 2008). This evolution of public water services is typical of industrialized countries; it is not clear whether it will also be the destiny of countries with less-developed water systems (Barraqué and Zandaryaa 2010).

The essential condition for the intervention of private companies in providing potable water services is the existence of a grid with reliable and cheap sources of fresh water. Otherwise, such firms are reluctant to enter into contracts due to uncertainty about profits and the availability of raw materials. This issue is more important in mixed cases: those in which water is not abundantly supplied by an efficient grid, but a multi-scale storage system exists. In uncertain cases, local authorities are tempted to outsource the service to private companies, seen as either more efficient or highly specialized in the water supply field (Spronk 2010). The private company makes up for local political weakness and shortcomings in civic culture, as evidenced by Italy, where there is a close correlation between strong municipal utilities and social capital (see Cartocci and Vanelli 2008).

The construction of water grids and privatization of the agencies providing the service raise questions not only regarding the storage structures (dams, cisterns, tanks, bottles and so on), but also the ways water is stored to conserve it in an era of waste. The idea that public authorities at all levels have of water conservation is unclear because future scarcity tends not to be considered or envisaged. The possible conservation policies revolve around three measures:

- less water-demanding domestic appliances;
- reduction in all uses of tap water;
- separation of the high- and low-quality water circuits.

The last measure has important implications for water storage. Several distinct circuits can be imagined: (a) providing drinking water from bottles or tanks provided by private shops and public kiosks, (b) using tap water for cooking and personal hygiene, (c) using rainwater for toilets, washing clothes or objects, and watering vegetable plots, and (d) having no water circuit for other uses, for example watering lawns and flowerbeds. Storage can be useful for all four circuits. The market has responded abundantly to the demand for bottled water: it is a formidable billion-dollar industry.[12] Municipalities and utilities are starting

---

12 Annual spending on bottled water in the USA amounts to $11.8 billion; global sales revenue from bottled water is $60 billion; http://www.statisticbrain.com/bottled-water-statistics/ (accessed 19 April 2014).

to open up high-quality water dispensers (kiosks) in public places – though this measure is contested by social movements which want such quality to be assured for tap water as well. The intentions and the spending capacity of users are crucial factors for separated circuits and rainwater harvesting. User–utility cooperation is essential for the success of multi-circuit systems: users have to change their habits, and the utilities have to arrange a functional infrastructure. Also, rainwater harvesting, unless it simply relies on barrels for landscape watering, requires large investments and advanced skills – resources that both private building owners and utilities do not usually possess.

This brings us to the main actors – the utilities. Because potable water delivery has become a complex task, such bodies have changed from simple branches of the municipality to industrialized companies serving broad areas, and sometimes listed on the stock exchange. They are, of course, sensitive to the need to conserve water, but only in terms of dealing with pipeline damage, which leads to waste and disputes with users. They cannot be so concerned to conserve through water harvesting. The use of rainwater for toilets and washing or irrigation entails large decreases in revenue for water utilities. In general, all water conservation systems are seen as factors that decrease the company's revenues (Barraqué, Juuti and Katko 2010: 23).

Water storage at building level is an ambivalent case: it does not lead to decrease in the consumption, but only rationalization, with the advantage for the utility of having to provide water only at certain times of the day or week. As mentioned above, this is the situation in many developing countries. Storage at end-user level is a way to ensure a constant service in places with growing water demands and insufficient supply.

Nevertheless, utilities' policies are generally not supportive of building-level water storage. Throughout the industrialized world, water harvesting is almost always a matter for single households or building managers. Utilities are responsible for providing reliable water pressure, so they install water storage towers or large tanks near the source, but they rarely intervene with grants, suggestions and container installation at the end-of-pipe level. Later, we will note the same pattern of behaviour with regard to energy storage and house insulation measures (see Section 4.3).

The most virtuous providers, for example Seattle Public Utilities in the USA, have introduced water fee rebates for households that install stormwater mitigation systems and barrels connected to gutters to collect runoff water. Such utilities install vaults, rain gardens, permeable pavements and infiltration devices that provide water quality treatment and/or slow down stormwater flows from impervious surfaces. Their rebates also extend to measures to reduce water use for toilet cisterns, washers and sprinklers.

Therefore, water conservation measures exist and are promoted by utilities, partially against their own interest in selling as much water as possible. Nevertheless, the limitation of these policies is that they almost exclusively cover single residence units with gardens, which benefit from much of the harvested water. The measures do not apply to the houses' internal plumbing; they are simple adaptations to a residential arrangement conceived for connection to an abundant grid source. The alternative is the off-grid system: harvesting water for almost all domestic uses, and possibly buying bottled water to drink. This solution is very demanding because it relies on a water treatment device installed inside the building.

The middle solution – mixing the grid supply with harvested water – seems quite rare, although in semi-arid areas it is envisaged for commercial buildings: 'Water from large buildings and malls could provide clean (distilled) harvested water for industrial uses or for mixing with well water to meet pollution standards for potable water. Cities could pay owners for harvested water' (Axness and Ferrando 2011, slide 17). The mixed solution evidently requires a complex management system with major technical and legal implications. Small-scale potabilization devices must be very reliable and adhere to hygiene standards. Another version of the mixed solution is to provide a dual supply inside the building – one for drinking and cooking, the other for washing – leaving harvested water to serve external uses. This has the disadvantage of being rather expensive for both the utility and building owners. The off-grid solution – with modest reliance on commercially supplied bottled water – is clearly the simplest. But full renunciation of the grid seems too drastic and irrational, at least in densely populated areas.

In conclusion, a grid connection integrated with robust storage systems within residential and commercial buildings appears to be the best solution both in areas subject to permanent water shortage and those temporarily affected (see Figure 3.1). However, with this solution, water harvesting is limited to external uses. Most well-organized utilities are already encouraging this solution by using rebates to promote the voluntary adoption of such measures. The installation of large cisterns inside or outside the building to provide drinking and cooking water is still the preserve of a small minority of enthusiasts. Of course, this picture varies throughout the world. The initial threefold typology can be applied: in poor countries, rain water harvesting is generally recognized as good practice; its implementation is a mix of traditional habits and improved techniques, though mostly without adequate maintenance (Baguma, Loiskandl and Jung 2010). In emerging countries like China, rainwater harvesting has been always practised in rural areas, while in urban settings the above-mentioned difficulties of integration with an unreliable mains system again arise (Barron 2009).

In developed countries, where water mains were installed many decades ago, rainwater harvesting systems are generally used for external watering, but some countries exhibit a more dynamic scenario. On the one hand, Germany is leading the way in Europe, for a variety of reasons: favourable norms, a widespread green culture and the application of robust technological solutions to environmental problems (Environmental Protection Agency 2013: 19). On the other, according to some statistics, Australia is the country where domestic rainwater harvesting systems are most common:

> The prevalence of rainwater tanks as a source of water for Australian households continues to increase. Twenty six per cent of households used a rainwater tank as a source of water in 2010 compared with 19% of households in 2007 and 17% in 2004.
>
> (Australian Bureau of Statistics 2010: 3)

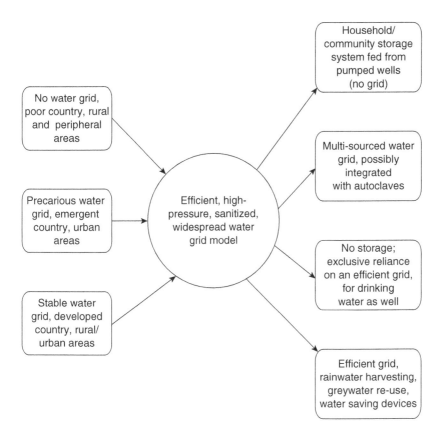

Figure 3.1 Evolution of potable water provision systems: points of departure, model adopted and outcomes according to the importance of types of storage

Certainly, recent periods of drought have stimulated favourable attitudes in this rich country.

Less comprehensible is the situation of south European countries, where these forms of water conservation have traditionally been common in both towns and the countryside. The reason is probably the type of infrastructural modernization undertaken in such areas. The main policy in southern Italy, as in Spain, relied on the construction of large reservoirs to provide water for major cities, often located on the coast (Barraqué and Zandaryaa 2010: 8). A major factor is the pace of infrastructure construction. Here is an instructive comparison between the United Kingdom and Spain:

> In the UK, the number of large dams grew rapidly during the 19th century from fewer than 10 to 175 at a rate of 1.7 per year. By 1950, the rate had almost doubled. After 1950, construction took place at a rate of 5.4 dams per year before slumping to zero by the late 1990's. Today, the UK has a total of

486 dams. By contrast, Spain saw the number of reservoirs grow at the rate of more than 4 per year between 1900 and 1950, before almost doubling and reaching 741 units by 1975. By 1990, this figure had more than doubled again (19.5 per year). Today, there are 1172 large dams.[13]

The slow rate of dam construction in Spain, as well as southern Italy, was due to delays in urbanization, especially of coastal areas, coupled later with the development of water-demanding beach tourism. Once adopted, the policy was generally correct: it was a way to fulfil the local population's right to a certain amount of tap water every day. The problems arrived later: the first was that connection to mains water led to greater consumption; the second was that all the local small-scale water storage infrastructure was dismantled or abandoned. Household or public cisterns were considered an out-of-date legacy from the past; in the brief history of water reclamation in southern Europe, the grid became an alternative to storage.

However, over time the contrast became less sharp. Richer and more organized households installed electric pumps and sealed tanks in order to meet their increasing need for water. In fact, despite the presence of large upstream reservoirs, the mains pressure was chronically low, certainly insufficient for modern devices like showers and instant heaters. But differences compared to the past provision system remained evident: the mains system, whether or not coupled with purification systems, provoked the abandonment of old practices: rainwater harvesting, fountain provision and the re-use of grey water[14] – techniques that today, in a period of water austerity, are being adopted in the most progressive urban areas of northern Europe.

The final result for southern Europe is a precarious multi-source provision pattern consisting of a badly maintained water grid (high waste and low pressure), almost exclusive use of bottled water for drinking, and variable degrees of self-provision in buildings, made possible by small-scale purification plants and internal tap water filters.[15] Thus, internal storage, previously systematically adopted by hotels and restaurants, is used to compensate for scant supplies of poor-quality potable water. This is a sort of forced autonomy – a contradiction in terms that can be interpreted in a number of ways.

13 http://www.eea.europa.eu/themes/water/european-waters/reservoirs-and-dams (accessed 24 April 2014).
14 Grey water is waste water produced by the washing of bodies and items. It should not be confused with black water (sewage) resulting from the use of toilets. The collection and treatment of sewage is still a problem in many rural and urban areas of southern and eastern Europe; http://www.eea.europa.eu/data-and-maps/indicators/urban-waste-water-treatment/urban-waste-water-treatment-assessment-3 (accessed 10 July 2015).
15 'In 2011, 9.3% of households in Italy complained about the irregular supply of water. The problem was most frequently reported by households in the South (17.4%), particularly in Calabria (31.7%) and Sicily (27.3%). Reluctance to drink tap water is still high in Italy: in 2011, 30.0% of households had one or more members who declared that they did not trust tap water. The distrust was highest in Sicily (60.1%), Sardinia (53.4%) and Calabria (47.7%)' (Istat 2012: 1; my translation).

According to the governmentality approach, recourse or appeal to the self-capacity of water users is a typical strategy of power. Public action on common problems is limited, while reliance on the abilities and knowledge of individuals is encouraged. The responsibility of the individual household is exalted, whereas that of the public authority is contained (Hellberg 2014). This is a view very distant from the classic interpretation based on the notions of *self-reliance*, *subsidiarity* and *voluntary commitment* (Bellah et al. 2007). Indeed, as we have seen, water harvesting, grey water re-use and the adoption of water conservation devices are presented as voluntary schemes dependent on individual choice. Furthermore, the rationality of the options is highlighted; these behaviours are rewarded by a reduction in water fees, together with undeniable advantages for the environment.

To *governmentality*, outlining the self-provision of vital resources as a manifestation of power, two other frameworks can be added *civicness*, according to which the adoption of water conservation measures is a manifestation of care for others and the environment (Morton and Weng 2009), and *instrumental rationality*, according to which all initiatives to store water are positive-sum games: utilities, users and the environment receive more than they give. These frameworks, based on relationships among interested actors, may have to be integrated with a factor concerning technological packages. Water storage has evolved tremendously in the past hundred years thanks to the construction of enormous dams, long pipelines and aseptic tanks. All these technologies have interacted with a range of organized groups, from the individual household to the multinational utility. The result is the dominance of a rigid model of provision – a single grid system – that 'makes it possible to adopt rationing systems able to discriminate among users according to their location but not according to essential and superfluous uses' (Ruggeri 2007: 8). Thus, the technological path adopted by modern towns a century ago can be changed only marginally, and only thanks to the users' goodwill. Various forms of water storage have roles in this effort, but users' goodwill is considered insufficient on the one hand, and risky (by absolving utilities of responsibility) on the other.

3.4 The big task of small farm ponds

Industrial uses of water are intrinsically linked to storage, almost always relying on an upstream high-capacity reservoir designated to feed a power plant or a cultivated field. Hydroelectric plants usually incorporate such reservoirs; those lacking any means of water accumulation, called 'fluent', are rare and less powerful, and more importantly, less controllable in a period when market prices dictate the planned provision of energy. This makes the use of reservoirs preferable: those supplying hydroelectric plants are refilled when energy prices are low and almost emptied when they are high.

The same applies to agriculture, whose pattern of development has followed that of industry: great amounts of energy must be available at the right moment in order to plan production efficiently. Thus, for agriculture water is the same as energy for industry. It enables increases in the scale of production. The abundant and regular provision of water comes essentially from high-capacity storage systems. Most of these rely on high elevations and deep valleys: a narrow valley is ideal for creating a reservoir. In these areas, settlements and agriculture are less common because of steep slopes and scant exposure to the sun, hence an artificial reservoir can be a good solution to valorize impervious terrain.

Of course, the history of great hydroelectric reservoirs has not been idyllic; in many places, they have been the cause of fierce (and unbalanced) conflicts (Barraqué and Zandaryaa 2010). They will be examined in Chapter 5 on biodiversity. This section will focus on small-scale water storage systems for productive use. These facilities are less well known than the Aswan High Dam, but yield numerous insights into the human capacity to conserve fresh water in good condition.

Farm ponds are an important example. They are small lakes, usually artificially created or adapted, used by farmers for irrigation during the dry season.[16] Some are more artificial, in the sense that the pond bottom is covered with a plastic film in order to retain more water in highly permeable soil. Miller (2009: 11ff.) identifies eight different types of farm pond systems. In any case, there is discontinuity with natural wetland because the water level of ponds is regulated to serve the purposes of farmers. If the objective is irrigation, farmers will almost empty these small reservoirs during periods of high water demand.

This discretion in water pond use is a key feature. Because of their small size and recognition of farmers' autonomy, ponds are generally subject to little or no regulation (Gustafson, Fleischer and Joelsson 2000), not least because they are on private land. They are in any case welcomed because their presence limits the farmers' tendency to devote as much land as possible to crops. Ponds promote environmental diversity – a process viewed favourably by both agronomists and ecologists – and improve farmers' image as professional land managers. Furthermore, especially in the global South, ponds can be used to provide edible fish as well as water for irrigation (Miller 2009). In these areas, diversification of the landscape enables the conservation of important plant and animal species that would otherwise be condemned to disappear through the industrialization of agriculture (León et al. 2010).

Ponds have several ecological functions (see Chapter 5). As well as conserving biodiversity, they retain nitrates that would otherwise pollute potable and recreational waters. Furthermore, they can reduce the flow of water during heavy rain – relating to flood prevention, as discussed earlier – and serve as a reserve in case of fire outbreaks. Their environmental functions include microclimate

---

16 The sizes and numbers of ponds are very variable: in Scotland, a country indeed rich in fresh water, there are 150,700 ponds, pools and lochans of up to 2 ha, and 4,500 lochs larger than 2 ha; http://adlib.everysite.co.uk/adlib/defra/content.aspx?id=000IL3890W.17UT2FLEXCG387 (accessed 2 May 2014). In Italy, farm ponds number around 8,000. They 'have a mean volume of 30,000 m^3 with some differences between North and South. In the North of Italy the average size of a farm pond is 24,000 m^3, while in the South it is 84,000 m^3' (Natali et al. 2009: 59). For further information, see Globevnik and Kirn (2009).

enhancement: 'water bodies capture and store solar energy and release this heat slowly, especially in the autumn, to the adjacent area' (Falk 2013: 98).

Finally, ponds can serve recreational and cultural purposes. Ponds close to cities can be used for angling, with a fee usually being charged, especially in the case of small ponds, which can be easily fenced. They may thus become 'club goods': a label to be applied to ponds whose use is restricted to members. The best golf club courses usually incorporate a small lake. In any case, stretches of water are admirable features of the landscape, aesthetically improving an environment flattened by industrialized agriculture. The symbolic meaning of ponds can lead to their being seen as sources of value: for example, a non-governmental organization (NGO) ambitiously adopted the name WATER for the WORLD and PONDS for PEACE.

Ponds are undoubtedly bound up with *civicness*. Their capacity to store water, especially in areas where it is scarce, means that there needs to be a stable agreement among users – in other words, peaceful coexistence. Farm ponds are consequently the focus of feverish political activity, and the unanimity about their usefulness conceals many unresolved problems. The first is the healthiness of stagnant stretches of water. As mentioned earlier, for hundreds of years people have been combating wetlands, often seen as sources of disease. The idea of creating numerous ponds around residential areas will not be happily accepted by everybody. Consequently, the NIMBY syndrome may arise at any moment. An important feature of ponds is that they contain stagnant water; if the water runs too quickly, it jeopardizes the pond's capacity to filter out pollutants and improve biodiversity. At the same time, mosquitoes and other vectors of disease may proliferate when water is too torpid. One remedy is the introduction of species that predate on such insects, like some types of fish and frogs. In any case, ecological balance is not easy to achieve at a time of climate change and the uncontrolled surreptitious introduction of foreign species into rivers and lakes.

The second problem concerns the balance of farm pond uses. It seems that exclusive utilization as a water reserve for irrigation may be too narrow in scope. The reason is simple: a single instance of use can lead to the complete emptying of a pond for a period that compromises the survival of animals and plants living in or around it. This is the same concept as *minimum flows and levels* (MFLs) for streams. If the MFL is not constantly assured – as sometimes happens with watercourses downstream from dams – the pond ecosystem may be completely destroyed.

We thus come to the third problem, which has already been mentioned: the regulation of small ponds located on farms, or at any rate on private property. The problem is only superficially different between poor countries with weak governments and rich countries with long traditions of regulating public goods. For both there is a problem of scale: finding the appropriate degree of integration between levels of water administration. Two cases illustrate this point. The first concerns the Apalachicola-Chattahoochee-Flint (ACF) watershed, an example of the long controversies concerning water use in the USA (Feldman 2008):

The new Florida lawsuit about Georgia's overuse of water in the Apalachicola-Chattahoochee-Flint (ACF) watershed makes a strong case that the problem is not just urban Atlanta. Agricultural irrigators in southwest Georgia also hold back and use immense quantities of water. Even worse, they tend to use the most water when the rivers are most dry: 'Over 20,000 non-federal water impoundments of various sizes have been constructed in the ACF Basin in Georgia. These impoundments intercept flow which would otherwise discharge to the ACF river system. The cumulative impact of these impoundments is significant, particularly during dry periods' [. . .]. Most of these impoundments in Georgia are 'farm ponds'. The litigators on Georgia's side of the dispute are aware that Florida water law provides a full regulatory exemption for most excavated farm ponds: 'Nothing in this part, or in any rule, regulation, or order adopted pursuant to this part, applies to construction, alteration, operation, or maintenance of any wholly owned, manmade excavated farm ponds, as defined in s. 403.927, constructed entirely in uplands. Alteration or maintenance may not involve any work to connect the farm pond to, or expand the farm pond into, other wetlands or other surface waters. This exemption does not apply to any farm pond that covers an area greater than 15 acres and has an average depth greater than 15 feet, or is less than 50 feet from any wetlands (section 373.406(13), F.S.).' The Northwest Florida Water Management District even publishes a brochure advising of the 'Benefits of the Upland Dug Pond' and how 'No permit is required from the District, nor most likely from any other federal, state or local agency.' Despite this encouragement, the Florida part of the ACF basin has relatively few farms and farm ponds. Almost all of the water intercepted on its path to Apalachicola Bay is held back upstream of the state line. Nonetheless, this farm pond exemption is a giant regulatory hole in Florida water policy.

This text, from a blog post,[17] illustrates the point very well: farm ponds are acknowledged as exempt from regulation even by administrations without direct jurisdiction over watercourses. According to the author, this is a 'giant regulatory hole'. Interstate watersheds are always difficult to manage, and federal unions are no exception (Hall 2006; Craig 2010). The problem is exacerbated by the very nature of ponds: small, numerous and situated on private property – in sum, difficult to locate, monitor and regulate (Globevnik and Kirn 2009: 19). In any case, the reason for their exemption is probably linked to the favourable bias toward small-scale agricultural initiatives in land use. Planning regulations covering the construction of farmers' homes and barns are usually less rigid if they are used exclusively for direct agricultural activity. Ponds are considered in the same manner. Moreover, ponds charged with ecological significance receive even more support from experts in various disciplines.

---

17 Tom Swihart, 'Farm ponds', *Watery Foundation*, 9 October 2013; http://www.wateryfoundation.com/?p=9173 (accessed 30 April 2014).

The EU Common Agricultural Policy equates ponds with other environmental adaptations, such as leaving field margins uncultivated, and planting trees and hedges. The maintenance and installation of all these landscape features are subsidized. In fact, these features have become obligations under the reformed CAP:

> Farmers with more than 15 ha of arable land will be required to dedicate 5 per cent of their farmland in 2015 to the creation of ecological focus areas. This could rise to 7 per cent by 2017 depending on the outcome of a review into their effectiveness. Field margins, hedges, woodland, fallow land, landscape features, buffers strips and ponds can be counted as Ecological Focus Areas.
> (National Assembly of Wales 2013: 2)

The second case concerns the state of Orissa in India, whose government has included ponds in its water and development policies. Again a blog post raises important issues:

> Dear Friends, if you notice, the Government of Odisha [or Orissa] has been presenting in the Public Domain various Policies on Development. In Odisha, we have such a poor Civil Society Network that there is no way to monitor how these policy decisions are translated into action and whether they have been beneficial or not. Of course, there is periodic analysis of the Budgets and Expenditures done very professionally by Pravash Mishra and Team in CYSD producing very authentic reports. Hope this exercise has some impact somewhere. Our common people do not have much understanding of the Budgets and Expenditures and hence there is very little response. In order that public mind is agitated over policy implications, there should be search for alternative ways of communication. Some people are happily indulging in Social Audits being hired by the Govt. and very ably doctoring this Post Mortem exercise. Very few are calling Spade a Spade.
>
> I am always cross-checking about the Policy declarations on construction of 1lakh Farm Ponds and Thousands of Check dams. Has the Government come with an open and verifiable statement how many farm ponds have been constructed under NREGS out of the target of 1lakh and what is the reality check is not known to anyone. Barring a few, I saw pits dug in the name of Farm Ponds causing sheer waste of public money under NREGS. Wherever I am going, I am also examining the check-dams constructed and their feasibility. Farms Ponds have very little Budget while Check dams have huge. In order to achieve targets, I am afraid, check-dams are becoming real jokes. So in both cases, there is fear that there are colossal waste of public money and defeat of the objectives of such measures.
>
> Our friends involved in RTI campaigns, must ask for the detailed report on Farm Ponds and Check Dams at the Block, District and the State Level and verify as many locations as possible. This will have two objectives – a) Make

Public Aware of the Public Policies and b) Improve the methods of implementation of the Policies in specific contexts.[18]

This text, written by the director of the Indian NGO Agragamee,[19] is exemplary of what the social movement literature calls an *advocacy task*. The author outlines three important repertoires of action: analysis of politicians' rhetoric on farm ponds, monitoring what is really done, and demanding account and improvement at all levels of administration. Expressions of support for farm ponds are probably typical electoral promises – ponds are concrete and comprehensible objects – but fulfilment of these is difficult to verify. Agragamee's mission is to close the gap between politicians and the common people. The NGO's director insists on increasing awareness and the intellectual involvement of poor people. There is no reference to *castes*, as previously made by Martínez Alier (2009), but to *tribes*, which are underprivileged rural communities labelled according to ethnicity. The author's evaluation of pond policy is rather severe, whereas other briefings on the same case are more neutral. The DHAN Foundation (2012: 14) underlines the general success of the pond policy in Orissa because the 'livelihood component has also been included' and public financial support has been scaled according to the size of farm: 'The small and marginal farmers will avail the facility free of cost while the other farmers will have to contribute 50% of the cost of such ponds to the Watershed Association.'

The DHAN Foundation (2012: 15) is more critical of the pond policy adopted in Tamil Nadu, the state where it operates. It underlines the difficulty of providing information to farmers about incentives and loans for the construction of such small reservoirs. Evidently, critical participation is linked to maintaining a visible presence in the area, a sort of *entrenchment* of third-sector organizations. Indirectly, the Foundation indicates the negative effects of the lack of a body like the Watershed Association, which in Orissa has probably been a good mediator between farmers and the authorities.

Some conclusions on water storage using farm ponds can now be drawn:

1   There are substantially two types of ponds: the farming version is stressed in the global South, whereas the multitasking one is more emphasized in the North. In the former, they are seen as providing livelihoods supplying water for consumption, irrigation and aquaculture. In case of the latter, much closer attention is paid to environmental performance. *Eco-services*, adequately rewarded, serve as the point of mediation between the two types (Massarutto and de Carli 2014).

---

18 Achyut Das, 'Policy utterances and reality checks regarding farm ponds and check dams', *FRIENDSOFKASHIPUR*, 21 June 2012; http://friendsofkashipur.blogspot.co.uk/2012_06_01_archive.html (accessed 27 January 2016).
19 Agragamee, which means 'pioneer', is a group of activists and thinkers committed to working with marginalized and underprivileged communities in the tribal districts of Orissa in India; http://www.agragamee.org/OldWebsite/aboutus.htm (accessed 27 January 2016).

2   Aside from geographical distinctions, farm ponds receive universal approval. Promising cases exist everywhere (Huang et al. 2012; Kakade et al. 2002; Verweij 2001). An agricultural pond serves as a valuable human–environment mediation point recognized by all. As a *boundary object* (Carroll 2012), it is at the centre of intense political activity not only by public authorities, but also NGOs and international institutions. Promotion and protection measures covering this multifunctional water storage system can be found everywhere.
3   We thus come to the problem of scale mentioned earlier: farm ponds are so universally sustained because they are numerous, small and independent – in other words, they reflect the beloved 'small is beautiful' principle. Indeed, water is difficult to manage at a local level because upstream–downstream conflicts arise very frequently; water storage is part of this precarious relationship, as the case of the ACF watershed demonstrates. Even in flat areas where the upstream–downstream dynamic is less evident, individual storage initiatives need to be coordinated, or better, 'reciprocally connected', which means very flexible storage of water. The creation of a *grid among ponds* would provide the answer to these problems of scale and coordination, and ultimately, that of water use privatization.

3.5 In between network and storage

Storage is not merely the simple action of keeping water in a container; it is almost always connected to other actions forming a practice, a bundle with the environmental conditions and the technological packages locally adopted for water delivery. It results in so many combinations of actions, conditions and instruments that an attempt at synthesis is necessary. Three main solutions have been envisaged in this chapter:

- self-sufficiency – off-grid, with great efforts to develop storage capacity;
- in-grid – complete reliance on a grid, with minimal or no storage at all;
- a mix of grid connection and robust self-storage capacity (see Table 3.2); this is valid for any water use, including ensuring water security.

Table 3.2 Relationship between types of water connection and water domains

Types	Off-grid	In-grid	Grid and storage mix
Water security	Floating or waterproofed houses	Higher and smoother banks	Partial self-collection of storm water
Drinkable water	Self-provision, purchase on the market	Pressure on the utility for service improvement	In-grid integrated with building cisterns
Industrial uses	Wells, isolated farm ponds	Full reliance on the irrigation scheme	Network of household or community ponds

The wholly off-grid solutions are almost always dictated by external conditions due to the long distances from organized water sources. They are rarely deliberate choices motivated ideologically, such as by a quest for an alternative lifestyle. Connection to a grid is the dominant and desirable system everywhere. It is the overarching preference justified on health and economic grounds: a grid to provide potable water and dispose of waste water ensures better hygiene, greater comfort, and often cheaper services. On the other hand, off-grid solutions relying on water harvesting and waste water purification using means such as reedbeds relies on large dedicated areas. Moreover, self-provision often requires the securing of authorizations, which may be laborious and expensive. It is nevertheless a practicable solution, to which the market responds by making a wide range of items and services available. The water container industry provides receptacles of all sizes, from 200 cl bottles to cisterns of thousands of litres. Services for drilling wells and equipping them with pumps are easily found, and are the typical forms of cooperation action in poor rural areas. Recourse to an in-grid mix plus large storage capacity arrangement is also made possible by the availability of this diversified set of items on the market.

The threefold typology also works well for water security. Again, single isolated houses with large open areas around them can organize for self-protection from floods. The simplest self-made solution is the digging of ditches and reservoirs around the building; another solution is leaving low ground empty or making it waterproof. Building a floating house is a further solution, though a very specialized one. Mass conversion to floating buildings would be very expensive. Improving a widespread drainage system external to privately owned land, which corresponds to the 'in-grid' in our framework, has been the most frequent solution, but it has proved too rigid in the case of heavy rainfall. For this reason, the grid can be helped by stormwater harvesting at house or block level. Detention basins with capacities of millions of cubic metres are the solution most often adopted; however, they are rarely connected to each other because this requires large investments and a great deal of land.

The threefold typology allows us to make some observations about policy actions. Relying only on grids, trusting the utility to supply potable water of higher quality and pressure, creates a sharp separation between demanders and suppliers: a situation labelled 'market exchange' if the former are really free to choose the latter. In fact, market relationships are realized only partially: there is and must be only one water grid supplier (a natural monopoly). The utility may provide a better service and the transparent metering of water flow. This *transparency* of costs and flows becomes the hallmark of a market in-grid water service. The value of transparency imposes a fee precisely linked to consumption and provides exhaustive information to the customer. Users must pay for the precise quantity of water they consume and waste. Thus, the utility adapts its strategy to a corporate model able to satisfy the demanding customer's preferences (*customization*). The fee is no longer a sort of tax paid to the municipal agency, but it must cover all the service costs, including the remuneration of capital invested by shareholders, who may be private financers as well as publicly owned companies.

This market-oriented policy is undoubtedly a source of political conflicts throughout the world (Swyngedouw 2005). The contentious issue is the reduction of water to a commodity, whereas it should be a common good. The criticism centres on two aspects: one is the idea that *calculation* is a precise way to establish the value of profit from water: 'in fact, water storage concerns usually the building of big dams, the only perceived way to assure stability of provision and hence stability of revenues for the organization enrolled in the water service, public or private it may be' (Hannigan 2006: 60). The other criticism is more subtle: it maintains that precise metering is a way to *control* people, who spontaneously assume a governmentality frame, based on the idea that:

> procedures, institutions, and legal forms [. . .] compel local people to internalize ways of knowing and conduct that reproduce and extend the influence and constitutive force of neoliberal doctrines – marketized regulation, participation, and rational (science-based) decisionmaking – within and between local socio-natures.
>
> (Ward 2013: 93)

The mixed situation – in-grid plus improved capacity for water self-retention – seems biased toward market solutions based on utility privatization, full fee costs recovery and the granting of profits. In fact, self-retention entails the precise metering of the water received from the grid and the water collected independently. This detailed procedure is necessary for calculating both the fee reduction and the residential unit's planned water needs. The needs of both demander and supplier for precise measurement suggests that the mixed type fully pertains to the governmentality frame: clients, wishing to control precisely their own consumption of water, voluntarily accept an order, a discipline, a cognitive frame. The order is incorporated into the household's self-monitoring. This theory interprets the search for autonomy as the embodiment of new, more sophisticated rules (Jones and Macdonald 2007). This happens at all levels – from the individual, through the household, to the community as a whole. The attitude to water service NGOs is symptomatic of this: they are seen as unconscious agents of eco-governmentality (Bryant 2002; Sending and Neumann 2006; Hellberg 2014; Empinotti 2007).

A relational perspective (Mische 2011; Donati 2011) furnishes a different interpretation of the mixed situation. This arrangement creates the conditions for an *enlarged exchange* between clients and the grid supplier (Osti 2012b): a broader agreement between users and the water source controllers. Users with powerful water self-storage systems can be valuable allies of a utility at times of peak demand for water. The utility may recognize this storage service with a fee discount. The same may happen in the case of stormwater self-absorption arrangements. Furthermore, the utility can reward the self-contained client by other means: for example, by providing grants for, or assistance with, the installation of home-level containers. This can happen only if the utility does not have an exclusively commercial aim, but instead a far-seeing management logic which takes into account user loyalty and uncertain water availability. An *enlarged exchange*, a sort of social pact, also arises when the utility provides more than one service (multi-utility) and the clients are its owners or shareholders (Mori 2013).

Hence, the supplier–client relationship is ambivalent: the mix of grid and water self-provision can exist completely within a neoliberal frame, or it may be modulated by social bonds able to generate mutual adjustments between water users and the utility. Self-storage may be the logical consequence of an isolated eco-entrepreneur's determination to be coherent in her or his beliefs. But the same actor may be part of a thick web of social ties that 'impose' the adoption of mutual help schemes, as in the case of a network of ponds or cisterns created to support households in the event of water scarcity or flood risk.

In conclusion, water self-storage of buildings connected to the grid – the mixed model – is a situation where mutuality is fundamental, but problematic. Each unit of water consumption is more independent, potentially self-sufficient; but the grid connection allows users to act reciprocally. This solidarity is more likely when the grid itself is a shared property. Thus, the final representation of the mixed model is a variable combination of three factors:

- the presence of a *third party* which manages the grid – an agency owned by the municipality or by users;
- *self-provision* – the will and capacity of individual users to adopt water storage measures at home;
- the *arrangements* between users and the agency considering mutual water needs.

However, centralized, large-scale and strict utility–customer exchange is the norm. It is a pattern lacking either self-provision or mutual arrangement. But if self-storage capacity grows, three changes are possible: (1) the utility stipulates specific agreements according to the storage capacity of users, (2) users in a particular area or block may find it opportune to reach a common agreement on levels of water consumption and disposal in order to avoid peaks of scarcity or abundance, and (3) a reduction of water consumption and management costs. This resembles a *smart water system* requiring not only a plurality of remote-controlled metering points, but also the material possibility of mutual exchange of water among individual users. Indeed, this is the essential meaning of a *network*: the conjunction of actors' independence and dependency, which in the case of water means combining self-storage with the capacity to supply other users. This can take the form of an immaterial network consisting of information exchanges and/or a material web of channels or pipes connecting cisterns, ponds and dams. This is the full meaning of an *enlarged exchange*, and provides a way to escape theories obsessed with control.

4 Energy storage

'The energy storage market is about to explode!' This was the slogan of the seventh annual storage week organized in February 2014 in Santa Clara, California. The slogan represents in iconic manner the current state of the energy storage system: it is an effervescent field full of very active operators who believe that energy storage devices and procedures are highly promising because they respond to revolutionary changes in the energy sector. Moreover, the slogan includes the word 'market', to indicate that the brilliant prospects for development are mainly connected to private exchanges among suppliers and demanders. The role of the public authorities is important in storage as well the entire energy sector, but the perception is one of new opportunities, mainly for companies.

In this chapter, these feverish storage initiatives will be unravelled from a sociological perspective. Energy storage is in fact a complex phenomenon which cannot be restricted to one industrial sector or to a consolidated field of applied research. It can be better described as a *socio-technical network*. Furthermore, there is a special linkage with the growth of renewable sources of energy: their intermittence requires complementary use of storage devices. Hydro power – the most widespread renewable – usually needs an upstream accumulation basin. Finally, energy storage entails more leeway for the final consumer, which modifies the relationship with grid energy suppliers used to being able to drive the demand side with ease. At first glance, energy storage is a way to sever the link with a network; in fact, off-grid storage systems allow self-sufficient provision of energy for a building, a block or an island. Indeed, the most interesting situations are where there is a mix of energy self-provision, storage and exchange with a grid, in similar ways to food and water storage.

4.1 The issue

Technically, an 'energy storage system [ESS] means commercially available technology that is capable of absorbing energy, storing it for a period of time, and thereafter dispatching the energy' (Malashenko et al. 2012: 4). This definition has three analytical components: the temporal phases, storage of energy using mechanical, chemical or thermal processes, and availability on the market, which means public accessibility as long as there is money to purchase energy. The market rule also means the possibility of legal possession of storage devices by private subjects. This last point is important in terms of relationships: whereas energy provision and connection imply compliance with a wide array of public rules, energy storage is a freer activity that underlines the final user's autonomy.

There are numerous techniques for storing energy (see Table 4.1), although there is a hierarchy in terms of use and maturity. There are natural settings and artificial arrangements in which energy is present in *potential form* or is expressed in terms of movement/work; the storage of energy refers to the former. The processes for storing energy are usually classified according to how materials or combinations of them can be used to accumulate and then release energy. Hence there are *mechanical methods* able to discharge movement when necessary, *thermal methods* able to release heat or cold when needed, and *chemical methods* with which to isolate substances able to supply thermal or kinetic energy at the appropriate time. *Electrical and electrochemical methods* use electricity processes combined with certain qualities of materials to accumulate and then deliver electrical energy directly. Finally, *biological methods* concern accumulation of potential energy in plant and animal tissues to be used for growth and movement. In this sense, food is also a way to store energy.

The complementary role of energy storage with regard to energy use emerges clearly from history: the first forms of accumulation concerned the collection and stowing of wood, then of other more calorific substances like fat, coal or oil. The most common way to store energy in order to produce movement is the creation of a water basin. The collection of water at a higher elevation (mill ponds) was the typical method used to drive flour and saw mills. After centuries of water and wood storage, fossil fuels arrived. Their capacity to be conserved in good condition over long periods is the basis for their popularity. Of course, fossil fuels are highly

Table 4.1 Classification of energy storage methods

Mechanical	Thermal
Compressed air energy storage (CAES)	Brick storage heater
Flywheel energy storage	Eutectic system
Gravitational potential energy	Ice storage
Hydraulic accumulator	Molten salt
Liquid nitrogen	Phase change material
Biological	Solar pond
Glycogen (body tissues)	Steam accumulator
Starch (staple foods)	**Chemical**
Electrical	Biofuels
Capacitor	Hydrated salts
Superconducting magnetic energy storage	Hydrogen
Electrochemical	Hydrogen peroxide
Flow battery	Power to gas
Rechargeable battery	Vanadium pentoxide
Supercapacitor	

Source: https://en.wikipedia.org/wiki/Energy_storage (accessed 31 July 2015).

flammable, but with some precautions, petrol or gas or furnace oil can be stored for a long time while keeping their energy potential intact. They also suffer from extreme temperatures less than other calorific products.

The ease of storage of fossil fuels explains their widespread use for automobiles. In fact, the real problem with electric cars is not the engine, which is much more efficient than thermal ones, but the batteries, which must store high potential energy in a small space and in various usage conditions. The main difficulty with power – which is more precisely a carrier of energy – is how to store it. It must either be consumed immediately after its production or conserved in heavy devices called accumulators or batteries. Other ways to store electricity include capacitors and hydrogen cells. Thus, the history of energy storage can be summarized into three fundamental stages (see Table 4.2): a long, quite static pre-industrial period; 200 years of the Industrial Revolution, and recent decades, when the post-carbon energy transition has begun (see McLarnon and Cairns 1989).

The industrial phase saw an important development: the capacity to exploit thermal energy to produce kinetic energy. This is symbolized by the steam engine, but finds its best expression in internal combustion engines fuelled by semi-liquid fossil sources like hydrocarbons. These must therefore be considered the primary energy storage system of the industrial period. However, they have been heavily criticized for a number of well-known reasons: their supply is destined to be exhausted, they produce large quantities of CO_2, and they pollute the air despite great advances in refinery and abatement systems. Moreover, although their capacity to store energy is very high, they are not included among mature industrial storage systems. This may be a controversial decision, but it corresponds to the policy trends of the major energy institutions.[1] Hydrocarbons persist and serve an important storage function, but they are considered an outmoded solution that must be superseded. A similar argument applies to nuclear energy. Uranium is a formidable concentrate of energy, but it has a significant problem: its use is difficult to

Table 4.2 Evolution of energy storage systems according to their periods of major expansion

Pre-industrial	Wood, mill ponds, food, peat, bodies of animals and slaves, pit storage of ice
Industrial	Coal, hydrocarbons, hydroelectric basins
	Small batteries in cars, trains and off-grid appliances
Mature industrial	Pumped hydroelectric
	Many others at a less commercially advanced stage

---

1 The OECD/International Energy Agency *Technology Roadmap: Energy Storage* (OECD/IEA 2014) seldom mentions fossil sources, and only for complementary tasks: for example, for thermal power plants with the capacity to store by-product hot water. Indeed, the roadmap's foreword states that nuclear power, with other technologies, 'will all require widespread deployment if we are to sharply reduce greenhouse gas (GHG) emissions' (OECD/IEA 2014: 1).

control, unlike hydrocarbons, which are so adaptable that they can be used at all energy system scales, from motorcycles to cruise ships.

The storage system symbolizing the third phase (mature industrial) is hydro power, in particular pumped water storage. In such a power plant, a basin is connected to a turbine and generator via a pipe that allows precise regulation of the fall of water and, above all, enables it to be pumped back up to the storage basin for future use. This has advantages in coping with peaks in demand and higher power prices in the energy market. Since the liberalization of provision and the unbundling of the energy chain, pumped hydro power plants have become the dominant storage system.

Power is the point of convergence among various storage systems; also the more experimental forms of accumulation are focusing on electricity. The source may be fossil fuels, but the important factor is the power output. This energy carrier is considered much more promising than others such as hydrogen. As long as its problems of storage can be resolved, electricity is the energy form for the future, not least because it fits very well with renewable energy sources, the four main ones being upstream water basins, wind, sun and biomass. Other sources are for the moment classifiable as only pilot schemes, such as the exploitation of marine waves and tides. Geothermal energy is a source that is highly concentrated in locations where there is volcanic activity, and it requires large installations and sophisticated technologies. Among all these sources, whether mature or experimental, only biomass can have a hydrocarbon as its primary product. It is not by chance that most biomass plants immediately transform the gas produced into power through combustion.

However, the use of biomass for energy production is criticized because of competition with foodstuff production. The other main sources are subject to less disapproval, and photovoltaic (PV) cells are universally appreciated (Ipsos MORI 2011), the main criticism being the amount of land required to deploy large-scale PV panels. Dams and wind farms are widely criticized, but this mainly focuses on their environmental impact, certainly not the fact they primarily produce electrical energy. The greater sustainability of renewables combined with their intermittence produces a short circuit between them and the need to store their power output. Hydro power is less intermittent because of the ability to regulate flows from upstream water basins, but it is still intermittent to a certain extent because it is affected by levels of rainfall and seasonal trends.

The dominant tendency to privilege power production risks neglecting thermal storage, a technical package that is much older than electricity: 'Thermal storage systems are perhaps as old as civilisation itself,' writes Dinçer (2002: 1). Thermal storage was primarily embodied in the construction of buildings, whose requirements for insulation include the basic principle of conservation of heat or cold. Thus, thermal storage is incorporated into a building's orientation (with its main façade facing south), construction materials (for instance, wood in an excellent insulator, but presents fire risks), the positioning of a heater/boiler (generally at the centre) and the choice of width of windows (wider or narrower, according to the need for warmer or colder air). Obviously, these construction techniques vary according to latitude and altitude, the presence of strong winds or tempering lakes, and many other factors.

Heat conservation is the pressing problem at extreme latitudes, whereas providing cold conditions is the main consideration nearer the equator. Woodpiles have long been features of the landscape of temperate zones, and even today, in mountain and rural areas they are an integral part of house courtyards. The conservation of cold is more difficult. Underground tanks or caverns have been used to store cool water or ice – the latter especially appreciated for conserving and transporting food during the hot season. Limited but crucial use has been made of stored ice in hospitals for post-surgery applications and the preservation of drugs during storage.

The first impetus for the industrial use of thermal storage came with the development of the modern systems of heat and cold production, which saw the introduction of central heating systems and refrigerators in buildings respectively. The *provision* and *conservation* of heat and cold proceeded at different paces, the former being so powerful that the latter was neglected: when thermal systems were generally quite cheap, the need for energy conservation was of secondary importance. An example is provided by instant hot water heaters in houses. When there is good gas pressure and the burner is efficient, it is not necessary to store hot water in a boiler; it can be produced immediately on demand. In some respects, this is a less wasteful energy system, and heaters without boilers can be much smaller, and therefore more suitable for smaller apartments.

The same applies to cold conservation: gas compressors for refrigerators are very efficient and the power to drive them is cheap, so the ability to conserve cold over long periods is of secondary importance. Air-conditioned houses can be built with barely any insulation because the costs of cooling or heating rooms are quite low. Of course, boilers and refrigerators, however small they may be, must have a certain amount of insulation. Consequently, the industrial phase of thermal storage began alongside the industrial production of heat and cold.

However, the main spur for thermal storage came with the ecological crisis, when the magnitude of the waste of energy to condition buildings became evident. Insulation techniques became more important than the efficient production of thermal energy. 'More important' is an ideal appraisal. Every technological path has such great inertia that the production phase of thermal energy still maintains supremacy over the insulation one.

The reason lies not only in the path dependency of the energy production sector or the capacity of energy companies to lobby; it also relates to the ease of measuring the performance of the production phase compared to the conservation phase. There are very accurate meters to record the production of energy, whereas measurement of energy conservation is less precise. Of course, in a laboratory it is easy to test the capacity of an insulating material, but performing the same test in the field with a real container made of that material can be more difficult because of the influence of many contextual factors. Furthermore, thermal storage (though this consideration applies to all forms of storage) reveals its performance over time – a factor that dramatically increases the variability of the results. The energy saving companies know this very well. The dichotomy between energy production and conservation is very evident in the green economy: on the one hand, renewable sources have received a great deal of attention and grants; on the other, technological packages for energy saving are less supported. In Italy, for example, renewables have for many years received copious grants (a positive incentive), whereas incentives for thermal efficiency, at least in the residential sector, arrived somewhat later and in the form of tax relief (a negative incentive).

The development of storage techniques is not simply a case of substituting one device for another, but rather relies on a bundle of several technological paths with different capabilities. As mentioned above, the bundles converging on power storage have received most support, whereas thermal storage, even though it is older, has received less attention. Finally, the most common form of storage in the industrial era – fuels of fossil origin – is not discussed in programmes for the future. The idea of drastically reducing the extraction of oil and gas from natural underground deposits in order to maintain a precious reserve of energy for the future is rarely considered. The only restraint is worries about lower prices because of high availability in the market. In any case, the high demand for fossil fuel among emerging countries overrides any concern for reducing the rate of extraction.

We may now recapitulate the different forms of energy storage after this digression on the two main types: power and thermal. We can use the sixfold classification in Table 4.1 based on storage methods. The taxonomy is very complicated, and the classes are not always mutually exclusive. Among the examples of each type there are several specifications depending on how some technological processes are classified by the respective disciplines. It is not important at this stage to analyse such details, because the important distinction is between devices whose use is widespread or which are commercially viable, and experimental ones:

> Despite the opportunity offered by the energy storage systems to increased energy stability and reliability of the intermittent energy sources, in 2009 there were only four energy storage technologies (sodium–sulphur batteries, pumped hydro energy storage, compressed air energy storage, and thermal storage) with a globally installed capacity exceeding 100 MW.
>
> (Evans, Strezov and Evans 2012: 4,142)

According to the US Department of Energy (DoE 2013), storage projects in USA are very unbalanced in favour of pumped hydroelectric storage (Figure 4.1). This accounts for 95 per cent of the entire capacity, equal to 23.4 GW. The residual capacity is more equally divided among thermal storage, batteries and compressed air storage. Figure 4.1 shows the minimal contribution of flywheels – a promising technology restricted for the time being to very specialized contexts like racing cars. The statistics are limited to 'grid storage projects', those that are easily accountable. Single off-grid initiatives, especially smaller ones, are presumably not surveyed.

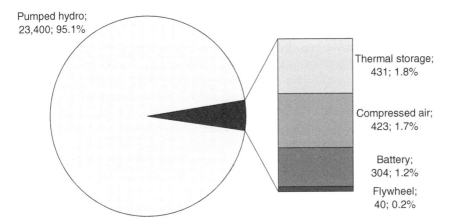

Figure 4.1 Rated MW power of US grid storage projects, including announced projects
Source: DoE (2013): 11.

Yet the sum capacity of isolated small cases of energy storage, even if numerous, cannot be compared with the scale of pumped hydro power. With regard to small storage plants, the first results of the incentives programme launched by the Federal Government of Germany are now available. In May 2013, the Federal Government approved a regulation supporting the development of a storage system through the German development bank KfW Group. A budget of €25 million was initially allocated to the programme (German Solar Industry Association 2013: 2). The KfW Group provides a €600/kW grant for new PV systems and a €660/kW grant for older systems with a capacity up to 30 kW. The subsidy must be used to reduce the interest rate on the loan contracted with the bank. The programme began very well: after ten months about 4,000 storage systems combined with PV panels had been installed. Because the subsidy is given only to PV systems of less than 30 kW capacity, we can calculate there is a maximum installed capacity of 120 MW. This is more than one-third of the entire capacity installed in the USA using battery technology (see Figure 4.1).

4.2 The legal, technological and social frames

Energy storage is located within broad and intricate legal and technological frames. The legal frame comprises the rules and authorizations applicable in each country; the technological frame relates to the numerous technological paths of storage devices. The broad classification adopted in the previous section is not universally accepted because numerous interests surround these technologies. These are not only economic – those of energy companies and device producers – but there are also different schools of thought[2] and concerns about energy security, which is a typical political issue.

The sociological approach adds a social frame – also called *embeddedness* – that helps to see interests, knowledge and legal norms as comprised of consolidated webs of relationships (Beamish et al. 2000). Such networks enable information and issues to flow, especially when the ties are weak, according to Granovetter's (1973) adage about the 'strength of weak ties'. Usually, a business community – like the storage energy system – can be conceived as a weak ties convention largely free of social barriers and prejudices. However, there are many social boundaries, principally national membership, a complex of traditions, attitudes and policies that create specific paths of development. A good example is the electricity grid: in Europe, despite economic and monetary integration, power grids are subject to stringent national barriers. Even in the USA, the champion of the free market, the grids are historically separated (Mazur 2013). The reasons for such barriers stem from a combination of technical and political factors that are generally no longer applicable today. But the socio-institutional embeddedness is not only a legacy from the past; it constantly takes new forms, for example that of a cartel of renewables companies which might prefer one storage technology over the others.

Another important frame that transects the ones previously mentioned is *scale*, a cognitive device which refers to both the physical density and the surface size of a phenomenon. It is the focal instrument of geographers, and is frequently adopted for the diffusion of innovation as well. The *transition model* (Geels and Schot 2007) is most frequently used today; it provides a three-level scale: niche-innovation, regime and landscape. It can aid understanding of the dynamics of energy storage, at least for conceptualizing at niche level, which corresponds to the application stage of many storage technologies.

In conclusion, the assemblage of all these frames provides a representation of energy storage as a *trajectory* (Rip and Kemp 1998) – a more or less rapid evolution of technological packages determined by company strategies, public regulation and incentives, and the fertilization capacity of open networks (Hauber and Ruppert-Winkel 2012).

Before applying this framework based on the idea of trajectory, we must add three technical aspects of energy storage:

- location in the energy chain;
- benefits stacking;
- suitability.

---

2 'There are two schools of thought regarding the use of advanced energy storage to integrate renewables. The first is that there is a huge total addressable market for advanced energy storage systems (ESS) when it comes to helping grids cope with variable generation – the challenge is finding the right business model. The second school of thought is that, although the market for integrating variable generation is sizable, ESS is being displaced by traditional power plants and the business case for ESS is actually quite challenging – not the least because the technology itself is still expensive' (Dehamna 2013).

Location is explained by the International Energy Agency (IEA) as follows: 'Energy storage deployment could be realised across the supply, transmission and distribution, and demand (end-use) portions of the energy system' (OECD/IEA 2014: 15). Accumulation is a process that can be inserted into every step of the energy provision chain: it is in fact a precondition for generation, like hydroelectric basins. It can intervene just after the energy collection, as happens with concentrated solar power plants, which accumulate the energy by means of molten salts. A storage device can be installed at the high-, medium- or low-voltage stages of a power grid. Finally, it can be placed near the end users, dividing this last location between 'community energy storage' and single-building level (Parra et al. 2015); this is undoubtedly the favoured level of research for the social sciences, especially now that energy self-provision is within the reach of consumers, who become *co-providers*, or *prosumers* (Juntunen 2014). This has increased the subjective actor-driven components of energy systems usually organized around rigid and centralized procedures.

The second technical aspect is known as *benefits stacking*: 'The ability for a technology or system to receive revenue from providing multiple compatible applications is referred to as "benefits-stacking" and is critical in the value proposition for many energy storage technologies' (OECD/IEA 2014: 12). Storage of hot water, for example, can be a good way to recover heat from a power plant; such water can be used to heat a whole district, but the system must be able to adjust the provision of hot water according to season. Combined heat and power (CHP) plants offer greatly increased efficiency and can reduce the fees to final consumers.

The multifunctionality of energy storage is not limited to the integration of heat with power. Within the electricity chain there are many compatible functions. Storage fulfils three main integrated purposes: achieving more efficiency/ reduction of waste, smoothing supply/demand peaks, and enhancing the grid's reliability and resiliency:

> Historically, storage technologies were predominantly installed as an investment that could take advantage of dispatchable supply resources and variable demand. Today, increasing emphasis on energy system decarbonisation has drawn awareness to the ability for storage technologies to increase resource use efficiency (e.g. using waste heat through thermal storage technologies) and to support increasing use of variable renewable energy supply resources.
>
> (OECD/IEA 2014: 6)

In the last instance, storage works as an efficient mediator between the variability of supply and demand.

The third technical aspect is *suitability*:

> The suitability of a particular technology for an individual application can be broadly evaluated in terms of technical potential. For electricity storage, discharge period, response time and power rating provide a good first

indicator on suitability. For thermal storage, storage output temperature and capacity can be used as a starting point in determining suitability for particular applications.

(OECD/IEA 2014: 13)

Translated into less formal terms, we have three main criteria. The first is time span – the speed and pace at which the energy is returned from the storage system to use or to the grid. In terms of pace of return, *power rating* measures the maximum amount of power a specific device can use or provide over a certain period (from seconds to months). Thus, a first distinction among electrical storage systems arises: on the one hand, pumped hydro power and compressed air energy storage (CAES) are not suitable for short-term responses to changes in grid demand; on the other, batteries, supercapacitors and flywheels can provide 'short-time applications, e.g. output smoothing within seconds, and applications on low voltage levels' (Zach, Auer and Lettner 2012: 8).

The second criterion is the number of possible recharges: in other words, charge/discharge cycles; this property is especially crucial for batteries. It is important not only for the calculation of economic costs, but also for lifecycle assessment. Evaluation of the entire life of a system allows estimation of its environmental and social costs. The factors usually considered for car batteries are their weight and autonomy, but not their lifespan, which is limited and greatly increases their cost.

The third criterion concerns suitability for the specific environment in which the ESSs are to be installed. Some are very large and cannot be installed in small apartments; others can tolerate only a restricted range of temperatures and need a protected or ventilated environment; yet others are noisy or potentially polluting (Ulu et al. 2013). Again, pumped hydro power storage systems are the most sensitive to environmental conditions; they must be located in depopulated areas without notable ecological value. For this reason, they have very narrow margins of growth in Europe, which is considered almost saturated, and new projects tend to provoke fierce conflicts in developing countries.

The study of energy storage trajectories requires the availability of datasets organized into series. The point of departure is therefore very demanding, whereas the situation is much more prosaic:

While some datasets exist that quantify the storage capabilities found in today's energy systems, attempts to comprehensively summarise the current global installed capacity for energy storage struggle from a lack of widespread and accessible data as well as conflicting definitions regarding what should be included in the baseline. Today, it is somewhat easier to establish a baseline for some countries, including the United States and Japan as well as some regions in Europe, for a specific subset of energy storage technologies. In these cases, data can be found for large-scale, grid-connected electricity storage systems. These data reveal that at least 140 gigawatts (GW) of large-scale energy storage is currently installed in electricity grids worldwide. The vast majority (99%) of this capacity is comprised of PSH [pumped storage hydro power] technologies [...]. The other 1% includes a mix of battery, CAES, flywheels, and hydrogen storage [...].

(OECD/IEA 2014: 16)

Thus, the imbalance in favour of pumped storage hydro power is even greater at international level than in the USA, according to the data presented above in Figure 4.1, which reflects the greater variety of storage applications in the USA. In any case, this is a socio-technical field whose trajectories are still in their initial phases. The IEA classifies the degree of maturity of energy technologies into three steps: (1) research and development, (2) demonstration and deployment, and (3) commercialization (see Figure 4.2). The classification can vary in terms of the number and qualification of steps (Schoenung and Hassenzahl 2003: 60, Table A.1; Chen et al. 2009: 306, Fig. 17). Some storage technologies can skip certain passages, but the general picture is clear: we have a technological archipelago – a large island (consolidated PSH) around which there are many other, much smaller trajectories of different forms (for an overview, see Luo et al. 2015).

In Figure 4.2, it is noteworthy that all the storage systems considered in the commercialization phase except PSH concern thermal energy.[3] The heat/cold

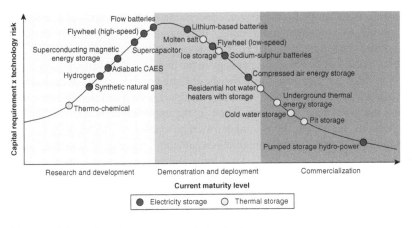

Figure 4.2 Maturity of energy storage technologies

Sources: Decourt and Debarre (2013); Paksoy (2013); OECD/IEA (2014).

---

3 Compressed air energy storage ranks second among the methods in terms of maturity; it is generally used for bulk energy storage. It compresses 'a gas (usually air) to high pressures (70 to 100+ Bar) and inject[s] it into either an underground structure (e.g. cavern, aquifer, or abandoned mine) or an above ground system of tanks or pipes to store energy. To generate electricity the gas is mixed with an additional fuel (e.g. natural gas), burned and expanded through a conventional or gas-fired turbine which runs a generator. Besides conventional CAES (also called *diabatic* CAES), there exists an advanced CAES concept, *adiabatic* CAES (AA-CAES), which does not need to use an additional (fossil) fuel to "restore" the stored energy (but need[s] an additional heat storage [system]) and also [has] a higher overall efficiency' (Zach, Auer and Lettner 2012: 8). It should be noted that CAES depends on hydrocarbons, and it generates CO_2.

storage technologies indicated are relatively simple, almost rudimentary, in that they are substantially based on the installation of high- or low-capacity containers. Thermal energy storage is not only incorporated into building construction technologies, but also rely on the use of water stored elsewhere, especially underground. Caverns are the ideal containers for water in its solid and fluid states; such cavities can work even at the gaseous phase, as in the case of geothermal energy sourcing from hot springs.

Thermal energy is the most widespread form in the world, the most exploited, and the most stored. Nevertheless, storage investments as well the attention of public opinion appear to be directed more at electrical energy. There is no conflict between the two forms of energy (and storage); on the contrary, there are many possibilities for integrating them. Nonetheless, the disadvantages of cold- and heat-based energy consist of two aspects: they need larger storage volumes, and they are less transportable. Despite many studies on how to address the shortcomings of thermal storage (Black and Strbac 2006), the point is that path dependency matters a great deal; thermal storage is a very old and traditionally less technologically dense trajectory.

Density is roughly a measure of the concentration of matter on a surface or a volume. In the case of technology, we can think of artefacts and the density of the relationships between them. Moreover, concentration of feedback and communication loops can be included in the concept of technologically dense environments (Bruni, Pinch and Schubert 2013). The degree of technological density can explain the position of each single package in the diagram of maturity presented in Figure 4.2. Whatever the position is, two aspects are crucial: one is the capacity of lay people (non-experts) to understand and manipulate such technology; the other concerns the possibility to include value in the application phase. The former aspect pertains to a concern consolidated within science and technology studies (Raymond et al. 2010); the latter requires explanation: some storage technologies show their value only after massive long-term application (the car battery is probably the best example). Density is thus a quality of the entire trajectory, and the main problems are in fact the uncertainties of the energy authorities and utilities that must choose and apply high-density storage packages.

4.3 The Tartar Steppe of energy storage

The Tartar Steppe is a novel by Dino Buzzati (1940) which recounts the story of an army sergeant, commanding a fortress on the Gobi Desert border, who spends his life waiting for an assault by the Tartar horde. The entire novel is a description of this wait for an enemy imagined to be as powerful as it is ferocious. Energy storage, with the sole exception of pumped heat electrical storage, seems quite similar. Everyone is certain that storage technologies will invade the market. Consultancy companies declare that the sector will grow four- to tenfold over the few next years (Lyons 2013). Thus, the innovation (the Tartars), with all its creative destruction capacity, will arrive. The developed world (the fortress) organizes conferences, meetings and fairs to prepare for the horde's arrival. But time passes;

every day the horizon is scrutinized, but only ambiguous signals are apparent . . . the final massive assault by the Tartars will not come before the end of the story.

The general consensus on energy storage revolves around a few applications, most of them at pilot level. According to the companies producing storage devices, including storage mechanisms and their connection to energy sources and uses, everything is ready for their massive application at every level of energy distribution and in every country, no matter what its stage of economic and infrastructural development. This means preparing for the commercial phase, where the storage packages can be bought at reasonable prices and installed everywhere.

Unfortunately, this is not the case, and it will be useful to summarize the possible reasons for this delay in market entry after the pilot stage of application:

- The costs are still very high compared to the inconveniencies of lacking the protection of the ESS: for example, a limited number of brief blackouts can be accepted either by users or providers; fluctuations in the grid due to intermittent sources can be tolerated. According to the eminent German consultancy centre EuPD Research, in 2014 power storage costs without public support schemes would still have been higher than electricity prices (Von Ammon and Pohl 2013: 4).
- The long-term functioning of the storage technologies in terms of maintenance and decommissioning costs are unclear (the lifecycle approach), as are the optimal working scales and combinations of storage devices. Experimentation takes a long time, and most storage technologies are still in their early stages. There are many uncertainties in evaluating the performance of energy storage systems (Ecofys 2014).
- The energy authorities do not send clear messages to energy providers and users about the usefulness of storage systems; they limit their mandatory prescriptions to many years ahead. They have not yet considered all the regulatory tasks entailed in multi-level use of energy storage (Bhatnagar et al. 2013). Financial support is also limited, and is provided only for individual projects. Furthermore, uncertainties over past subsidies for renewables make the authorities cautious, because they do not want to provoke further criticism of their policies.

These considerations are summarized as the points of a triangle (Figure 4.3).

Another explanation is external to the triangle of energy storage. It concerns who might be damaged by the introduction of massive energy storage devices at all levels of the chain. There are two main actors that stand to be negatively affected: energy producers that use fossil fuels, and the utilities. However, a subheading in a report on the energy storage situation in Germany declares: 'Conventional generation threatened, but opportunities in distribution and supply' (Dickens et al. 2014: 6). It is noteworthy that both conventional energy producers and the utilities also initially opposed renewable sources of energy (RSE). The traditional generation sector viewed RSE as an unfair competitor that used free sources like the sun and wind, often with assistance from the state; the distribution and supply sector

110 *Energy storage*

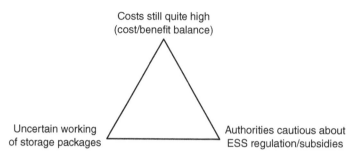

Figure 4.3 Reasons for low penetration of energy storage

also saw RSE as a competitor that would reduce clients' energy demand because of self-provision. In both cases, it is not only a matter of being able to sell less energy, but also of having to establish a new, more demanding relationship with energy users. When consumers are able to self-produce some of the energy they need, the one-way direction of the flow is interrupted and new terms of exchange must be established. This entails the development of *smart grids*: multi-source systems whose management will complicate the lives of both the utilities and the centralized energy producers. Both must become much more flexible in provision: thermal power plants fed with naphtha or coal have structural problems that impede such flexibility; the same applies to nuclear power stations. Gas turbine plants are better able to respond to rapid changes in demand for energy. Energy distribution companies can use the grid as a source of flexibility, but only within strict limits, and even sophisticated software can only manage the variability of a multi-source energy system within a limited range:

> Utilities have both financial and technical concerns about how microgrids will influence their business model and the functioning of the central grid. On the technical side, utilities worry that microgrids may harm the reliability of the larger grid through faulty interconnection, tripping or failing to island or re-connect correctly. On the financial side, utilities express concern about the cost to provide back-up power for microgrids, especially if they proliferate. [. . .] But much of the discussion centers around the 'utility death spiral,' the idea that customers will flee the system for distributed generation and microgrids in great numbers, leaving the utility with a rate base too limited to fund needed infrastructure without dramatic rate increases – which will in turn cause further customer flight.
>
> (Wood 2014)

If energy microgeneration causes such turbulence for fossil energy plants and utilities, the effects of massive inclusion of storage systems exacerbate the situation, but more ambivalently. In fact, on the one hand, when energy stored at the end user site is combined with an internal source, it leads to decreases in flow in the grid and demand on energy plants. The consequence is lower revenues from energy sales and grid hire.[4] On the other hand, the ability to store energy in many small sites increases the stability of the grid. Storage, in fact, smooths out the peaks and troughs of energy demand, making the entire system more predictable. Oscillation of demand and supply is less than in systems where there is no storage, but a massive presence of intermittent sources like renewables. The burden of variability management is shifted to every energy prosumer, who must acquire the means to control the consumption and storage of energy, and possibly export it to the grid.

This last point probably explains the reluctance of final energy users to adopt energy storage systems: they must bear the cost of managing the flows, which may be expensive, difficult and prone to incidents or rapid attrition. Furthermore, there are non-monetary costs: for example, the time required to learn about and monitor the storage system, even if it is highly computerized. A trade-off between automatic and discretionary behaviours arises (Carrosio 2015b) – a dilemma which discourages the final users. Contrary to common sense, final users are not so inclined to the ideal of self-sufficiency, preferring inclusion, or otherwise, active resistance (see Moisander and Pesonen 2002; Eräranta, Moisander and Pesonen 2009; Knox-Hayes et al. 2013). If the grid guarantees a minimal level of stability in energy provision, with reasonable costs, the consumer will prefer to trust the utility, avoiding too many complications at home. This digression on the demand side of the issue recalls the idea that storage is a source of high variability in the asymmetrical relationship between providers and final users.

Indeed, storage would weaken this asymmetry if only the consumers wanted it. The economic damage due to a decrease in sold and distributed energy may be of less importance for a rich and multitasking utility. What is at stake with the combined self-provision/storage system is the bargaining power of the final user. For prosumers equipped with an ESS, the asymmetry with the grid provider diminishes. This process is much more evident than in the case of renewables. If there is both an internal and external energy source, a net metering system is sufficient. This is a simple 'mechanical' device, because it only has to register entry and exit flows.[5] Instead, as mentioned above, storage needs intelligent management: at least a model of energy consumption must be programmed by the final user. This model can be negotiated with the energy grid provider, for example by charging different fees according to the time of day (time-of-use tariffs).

But storage is also part of a thick web of internal relationships within the energy sector. Here the situation is even more complicated, because the ownership

---

4 The liberalization of the energy provision market did not concern the grid, which remained the sole supplier; all authorized energy sellers must pay a fee to the grid owner for the energy they sell to their clients. The grid is a natural monopoly that reduces the reach of energy liberalization.
5 Net metering devices are electronic instruments, often incorporated in remote-control systems; but their basic function is mechanical. Consumption feedback services are necessary to enable net meters to fulfil social functions like energy equity (see: Podgornik et al. 2013).

of plants and grids is variable, especially in those countries that have moved more decisively towards unbundling. This is especially evident in the case of wind farm energy. In Italy, for example, ENEL, the former state power monopoly, owns wind farms and most of the country's low-voltage grid (distribution). It has a subsidiary for the promotion of renewable plants, including PV panels. The difficulty in balancing these interests within such a large company is evident. And it is a task that is even more problematic for energy companies operating in Germany, where the transition (*Energiewende*) is more advanced (Dickens et al. 2014).

The consequence is that there may be conflicts concerning energy storage within the same company. The core business of large energy companies will probably focus neither on renewables nor storage, and they will experiment with renewables only for symbolic purposes – to please public opinion superficially attracted by the idea of a green economy. Conversely, adopting a perspective based on internal socio-technical groups competing for different trajectories can provide other insights (Callon 1987). These concern an energy company's conflicts:

- with other companies providing the same kind of service, whether or not using the same technology – *market competition*;
- with companies using different energy sources and technological paths – *inter-sector competition*;
- within the same company among segments engaged with different sources and technologies – *internal competition*.

As regards the first conflict, it is noteworthy that a handful of multinational companies (including NEC, FIAMM and Siemens) dominate the high-voltage storage market, at least in Europe. At the smaller scale, especially for battery technology, the field is much more crowded and competitive in a market that is already globalized, and in which even small or medium-sized companies are able to compete.

Inter-sector conflict especially concerns the utilities. A key challenge in the wider adoption of energy storage is the standardization of procedures. In considering the electricity system, Ecofys (2014: 47) lists three main sources of resistance: (a) a high degree of caution among utilities in adopting new technologies, especially because of the intricacies of norms and regulations; (b) a lack of standardized controls and interfaces for energy storage with respect to existing utility energy management systems, and (c) cross-functional operability due to the unique nature of energy storage compared with other power system components. Clearly, the problem is *relational in nature*. It is a problem of connecting one technical subsystem with another, given that regulation is lacking or confusing.

This issue emerges in the following case, which concerns one of the most advanced places in the world in terms of storage:

> San Francisco – SolarCity Corp., the biggest developer of U.S. rooftop solar panels, halted efforts to install and connect systems that include batteries for power storage because California's utilities are reluctant to link them to the electric grid. About 500 SolarCity customers in the region have agreed to use

> the systems, and the state's three biggest utilities have connected 12 of them since 2011, said Will Craven, a spokesman for San Mateo, California-based SolarCity.
>
> SolarCity is testing the units with photovoltaic panels to generate power and batteries that retain that energy for use when the sun isn't shining. The combination makes customers less dependent on local utilities. It may be a threat to the business model that's underpinned the power industry for a century. 'We've stopped submitting applications because we've lost faith that these things are actually going to be carried out in any reasonable time,' Craven said in a phone interview.
>
> The utilities require a series of applications and fees that Craven said makes the process too onerous. SolarCity has installed a total of 65 of the systems in areas overseen by PG&E Corp., Edison International's Southern California Edison and Sempra Energy's San Diego Gas & Electric. 'The ones we have submitted haven't gone anywhere,' he said. Homeowners with rooftop panels buy less electricity from the grid, and those who use batteries to store power may need to purchase even less.
>
> (Doom 2014)

The utilities' reply is on the same Web page:

> The utilities say they support the use of solar power and new technologies such as batteries that promote energy efficiency. They also note that storage is a relatively new capability and that it will take some time to properly assess how to add it to the grid at fair pricing. David Eisenhauer, a PG&E spokesman, said it takes about eight to 10 weeks to handle applications and the utility has processed eight of the 20 it has received. "Because battery installation is such a new technology," he said today in a phone interview. 'We're still working to find more efficiencies in processing the applications.' San Diego Gas & Electric said there is 'an ambiguity in the existing tariff language' regarding storage and it's working with regulators to determine the appropriate fees, Hanan Eisenman, a spokesman, said in an e-mail today.
>
> (Doom 2014)

In public debate, utilities minimize the problem. They say that it is contingent, procedures will surely improve, and that there is absolutely no prejudice against energy storage. In fact, they face the same issues as reported at the beginning of the PV panel era. Utilities delayed installation, blaming the uncertainty of rules, inability of developers, and improper burdens on the grid managers. Such problems were overcome when rooftop PV panels spread widely in the countries that adopted subsidies. Then it will be only a matter of time: after the pioneering connections are in place – in fact, SolarCity talks of hundreds, not thousands, of storage installations – the situation will be normalized, with utilities in a position to 'grin and bear it'.

Nevertheless, the parallel with PV panels holds only with the introduction of subsidies; without them, a mature technology like PV panels would probably have remained confined to marginal uses for ships, islands and mountain huts. This calls for massive public intervention in favour of energy storage, especially at household level. The demand side still remains weak; in Italy, over a few months a spontaneous initiative launched by an energy blog in June 2013 collected 500 memberships of a 'purchase group' with the task of selecting the most convenient provider of a small ESS. The initiative died after a public statement by the Italian Gestore Servizi Elettrici (GSE) that declared the ending of incentives for PV plants to install storage systems (Meneghello 2013). The GSE called for official intervention by the Italian Energy Regulatory Authority to apply definitive regulation to the interchange between the grid and combined PV/accumulator systems. A year and half later, in December 2014, the Energy Regulatory Authority's clarification arrived, but it is premature to expect a mass diffusion of storage systems at household level even though such installations can benefit from tax relief (Pigni 2015).

4.4 Power storage systems embedded in national policies

In order to explore possible mechanisms for energy storage development, the delays in the Italian case can be compared with the pace of development in other areas: California, Germany and France. In the first two cases, their inclusion is immediately understandable: they are world leaders in intermittent renewables, and they are pioneers of legislation in favour of energy storage. France acts as a sort of counterbalance because of its scant diffusion of intermittent renewables and the dominant presence of a centralized (mostly nuclear) system of energy provision. Its situation could be, paradoxically, a major stimulus for the introduction of energy storage systems. It is well known, in fact, that nuclear power energy provision is very rigid – constant over time – and a calibrated system of storage could give it more flexibility. It could even turn France into a country that introduced massive ESSs without a transition through renewables. At this point, Italy would be a case where, despite the wide presence of renewables, energy storage is blocked by political factors concerning not only the uncertainty of governments, but also weak demand by final energy consumers. A third explanation concerns the role of intermediate actors like utilities and firms committed to the provision of energy devices. Their view of the issue and their lobbying actions become crucial factors in the implementation of programmes more or less favourable to ESSs.

Interesting Asian cases will be considered later; Japan and China are making large investments in renewable and energy storage systems (REN21 2014: 14). The initial exclusion of Asian countries from the analysis is justifiable to maintain a degree of uniformity according to a *comparative political economy approach* (Evans 1989; more specifically on energy, see Mitchell 2007): they are countries in which energy decision-making is dominated by a strong alliance between state agencies and large companies, according to a top-down 'directorial' model in which there is little room for local utilities and the demand side.[6] Like any model, the representation is over-simplified and risks missing important local factors of energy storage activation; for example, some countries, like Indonesia and the Philippines, have a complex political structure that can play an unexpectedly vital role in promoting decentralized energy systems (Marquardt 2014). In any case, countries in the Far East can serve as counterparts to a supposed Western model.

The analysis will be based on Table 4.3, which compares some crucial data among the four cases. The choice is to insist on intermittent renewable sources of energy because they appear to be the main triggers of change in storage practices. The sources are wind and PV solar panels; other sources, like geothermal energy, sea waves, hydro power and biomass, are more predictable, and in some cases they themselves serve as forms of storage. On selecting sun and wind sources, a further choice is made: to analyse only electricity generation, transmission and use.

California is the US state with the largest presence of solar renewable energy systems. Its 2012 production (1,834 GWh) was much less than that of the three European countries in Table 4.3. In proportion to inhabitants, California has a value similar to France (48 kWh per capita), while Germany and Italy reach more than 300 kWh per capita. By the way, in 2013 California had a huge increase, doubling its PV production (4,291 GWh). This state is better positioned for wind energy, producing a total of 240 kWh per capita in 2012, a value very close to that of France (227 kWh per capita) and Italy (225 kWh per capita). All three are very distant from Germany, which in 2012 produced 630 kWh per capita. Notably, in the same year German wind farms provided 8 per cent of its total power production. Among the four cases, France is the country with the lowest contribution from intermittent sources. We have excluded the important source concentrated solar power because it is still an experimental technological package deployed in only a few countries, the leader being Spain. It is itself a form of energy storage able to regulate heat from the sun, thus reducing the intermittency of this source.

In each country, the percentage of power covered by wind or PV panels is still quite low, usually below two digits. However, the intermittent sources continue to grow and to maintain priority in grid supply. Their importance will certainly increase, requiring new policies. There may be defensive actions by thermal power producers whose aim is to reduce the incentives for renewables, and which press for subsidies for loss of production/maintenance of traditional plants (known as *capacity mechanisms*; Crisp 2015). These backward policies may be successful in the short term, but they will be improvident in the long run.

---

6 'The [Asian] method emphasizes economies of scale, most clearly demonstrated by traditional "hard paths" of fossil fuel based grid connectivity. This process has mostly taken a top down approach with little feedback in planning stage leading to copying of external (western) models. The national level energy policy making has been mostly a central government enterprise leading to less socially acceptable decisions, absence of effective delivery institutions and less transparency leading to policy inattention or deliberate non-action' (TERI-IGES, AEI 2012: 17).

Table 4.3a Crucial factors for energy storage systems in four areas

	California	Germany	France	Italy
Production of intermittent renewables (GWh, 2012)	Wind: 9,152 (4.6% of in-state generation)[1] Solar: 1,834 (0.9% of in-state generation)[1]	Wind: 50,670 8.1%[2] Solar: 26,380 4.2%[2]	Wind: 14,913 2.7%[2] Solar: 4,473 0.8%[2]	Wind: 13,407 4.5%[2] Solar: 18,862 6.3%[2]
Total electricity produced 2012 (TWh)	199.8[1]	629.8[3]	560.7[3]	299.3[3]
Population 2012[4]	38,062,780	80,425,823	65,649,570	59,539,717
kWh per capita	Wind: 240 Solar: 48	Wind: 630 Solar: 328	Wind: 227 Solar: 68	Wind: 225 Solar: 317
The world's largest electric utilities as of 1 April 2014, based on market value (in US$ billion)[5]	Duke Energy: 44.4 (not based in California) California's three major utilities: PG&E, SDG&E and SCE	E.ON: 37.6 RWE Group: 25.1 (with EnBW and Vattenfall, they form the 'big four')	EDF: 75.5 GDF Suez: 64.6	ENEL: 53.2
Total capacity market share by top 15 energy storage technology vendors[6]	Gridflex, 13%(Boise, Idaho)	Voith, 16%	Alstom, 19%	—
No. of energy storage projects (non-pumped hydro power projects)[7]	132 (122)	48 (20)	25 (14)	35 (17)

Sources:
1 Total electricity system power without imports, http://energyalmanac.ca.gov/electricity/system power/2012 total system power.html (accessed 23 April 2015).
2 https://en.wikipedia.org/wiki/List_of_countries_by_electricity_production_from_renewable_sources (accessed 23 April 2015).
3 http://www.bp.com/content/dam/bp/excel/Energy-Economics/statistical-review-2014/BP-Statistical_Review_of_world_energy_2014_workbook.xlsx (accessed 23 April 2015).
4 http://cait.wri.org/ (accessed 8 February 2016).
5 http://www.statista.com/statistics/263424/the-largest-energy-utility-companies-worldwide-based-on-market-value/ (accessed 23 April 2015).
6 Navigant Research, *Energy Storage Tracker 3Q13*; http://www.navigantresearch.com/wp-content/uploads/2013/07/TR-EST-3Q13-Brochure.pdf (accessed 4 July 2014).
7 US Department of Energy Global Energy Storage Database; http://www.energystorageexchange.org/projects (accessed 4 July 2014).

Table 4.3b Crucial factors for energy storage systems in four areas (presence of a movement)

	California	Germany	France	Italy
Number of PV plants applications	91,235 applications installed under the CSI programme in 2007–2012 for a total of 341 MW capacity (96% residential)[1]	1,223,000[2] '98% connected to decentralized low-voltage grid'[3]	281,724 in 2012 Residential systems up to 3 kW accounted for 86% of all installations and 16% of total power capacity; systems over 250 kW are 0.3% of all installations and 44% of total capacity[4]	478,331 whose 88% is less than 20 kW peak (15% of total PV capacity)[5]

Note: The number of PV plants has been taken as indicator of the presence of an energy movement.

Sources:
1 'Installed applications, California Solar Initiative-CSI Programme, 2012–2012', http://www.californiasolarstatistics.ca.gov/reports/monthly_stats/ (accessed 14 April 2015).
2 Bundesverband Solarwirtschaft e.V., *Statistische Zahlen der deutschen Solarstrombranche (Photovoltaik)*, September 2012. http://www.solarwirtschaft.de/fileadmin/media/pdf/bsw_solar_fakten_pv.pdf (accessed 14 April 2015).
3 Wirth (2014): 37.
4 ADEME (2013).
5 Gestore Servizi Energetici (2012).

In this situation of conflict between a dominant energy industry based on fossil fuels (nuclear ones included) and a much smaller industry of renewables, energy storage can be a *solution for all*: (1) it can solve the problem of intermittency for providers of renewables; (2) it can provide a way to absorb the over-capacity of large fossil fuel energy companies because they can store more energy and sell it when they want, and (3) it can help transmission agencies to provide a more secure service. However, storage devices at these levels are very expensive, and it is likely that the companies will ask for public compensation for large investments in ESSs.

In any case, a major threat to traditional power providers remains, 'since a storage system can increase the percentage of self-consumed energy from 30–40% to 60–70%' (Codegnoni 2014). Even though the range is rather large, its minimum value is so high as to alarm any energy seller. A reduction of that extent can create real problems of survival even for giant, multitasking energy companies. For this reason, the larger energy plant companies and utilities have already moved to renewables for a certain quota of their portfolios. The other complementary strategy would be to sell storage services to clients, but in this case the companies would have to evaluate the costs of learning and operating in a new sector in comparison with simply continuing to sell energy. For all these reasons, the combination of small PV panels with storage remains a major challenge for the entire energy chain.

118 *Energy storage*

Wind energy is less challenging. It has developed mainly relying on large turbines and usually through investments by mid-size or large companies. For various reasons, small wind turbines are not as widespread as PV panels (Osti 2012b). Thus, the bulk of wind energy is channelled to the high-voltage grid without any opportunity for self-consumption. Of course, storage fits very well with the power generated by wind farms, but for reasons different from self-consumption. Storage makes it possible to avoid wasting wind power because of inadequate transmission lines, allowing compensation for peaks and troughs of production. Finally, the scale similarity between wind power generation and energy companies makes the prospect of storage less traumatic, if not a source of greater revenues for investors in the sector. If these investors are also the owners of thermal power plants, it is very likely that a conflict will arise within the organization. On the other hand, if wind farm owners are companies outside the electricity generation oligopoly, the conflict moves to other fields: business associations, sectors of government ministries, and the energy media.

For wind power, because of the great capacity of turbines, the main issue concerns the owners of high-tension transmission lines. In Italy, these are managed by a quasi-public company, Terna, defined by the European Union as a transmission system operator (TSO), which receives and supplies wind-generated power to the high voltage grid via its monopolistic structure. In Germany, there are four TSOs, each operating as a natural monopoly in several *Länder*. In France, there is the Réseau de Transport d'Électricité, a branch of EDF, a publicly owned company which is the national power monopoly and also the largest energy company in the world (see Table 4.3a). In California, an independent system operator 'manages the flow of electricity across the high-voltage, long-distance power lines that make up 80 percent of California's and a small part of Nevada's power grid'.[7] This operator, which is controlled by the Californian power utilities, has the task of ensuring open and equal access to sellers and buyers of energy – a commercial task that in other countries is undertaken by a public body. Then, among the four selected cases, there are slightly different formulas for dealing with high-voltage wind energy.

In terms of independence among the actors in the energy chain, these grid operators present two complications: one is ownership, which can be entirely public (in the French case; via the TSO TenneT in Germany), mixed with a golden share in the hands of the public (Italy), fully private (Germany) and even non-profit (California). In some cases, the distinction between TSOs and energy production companies is only formal; in practice, the various firms in the power chain belong to the same holding companies (Brunekreeft et al. 2014). The other complication concerns functions: usually such agencies provide transmission and balancing services in a monopoly regime. According to the promotional literature of one of them:

---

7 http://www.caiso.com/about/Pages/OurBusiness/UnderstandingtheISO/default.aspx (accessed 7 July 2014).

Transmission services: TenneT transmits electricity via the high-voltage grid in the Netherlands and large parts of Germany. We connect producers to the regional grids which supply electricity to consumers. System services: TenneT operates and develops systems that manage and maintain the balance between electricity supply and demand.[8]

In order to ensure balance in the electricity system, such companies are tempted not only to channel the flow, but also to feed it directly – for example, with a mixed and strong system of storage. This is an awkward aspect in those systems where the provision of energy is legally and operationally separated from transmission: so-called *ownership unbundling* (Brunekreeft et al. 2014: 19). In the case of massive use of storage, the grid transmission agency can be accused of invading the generation company's field of action.[9] This is not a problem in those cases where the unbundling is only legal and the various companies in the power chain cooperate rather than compete for the provision of services at various levels. It is evident that in a situation where there are fuzzy borders between functions (an imperfect market), the wind energy production/power storage combination becomes even more attractive because it responds to a need for the flexibility of a multi-source power system. Thus, there is generally no conflict in terms of storage between production and transmission bodies working in a regime of quasi-monopoly or legal unbundling.

In conclusion, what really creates a divide within renewable energy sources and storage systems is *size*. The scale of the renewable plant can be used to distinguish between two subsystems, as shown in Figure 4.4.

It is apparent that there are two separate sets of actors and actions in Figure 4.4. On one side, there are large-scale storage arrangements inserted at the transmission and distribution levels, on the other, small PV and battery systems installed in the buildings of final power consumers. In Germany, they coexist, in the sense that the small-scale storage system is somewhat stronger than in other countries due to the support of a grassroots movement in favour of solar power.

The question of whether a *social movement* for energy transition exists is not easy to answer. The strength of the German Green Party is well known, together with the country's cultural sensitivity to natural settings (Markham 2008). The German government's public prioritization of PV, followed by small-scale storage, can be interpreted as a response to these forces. The case that most closely resembles Germany is probably California, where a special combination of public awareness and pro-solar power choices by the state legislatures has produced the

---

8 http://www.tennet.eu/nl/about-tennet/organisation.html (accessed 20 April 2015).
9 For example, there is conflict in the USA over whether '[b]atteries act like generation resources, so they should remain part of the competitive market, which can better handle and appropriately price battery technology risks' (St. John 2014), or whether they are instead tools used to smooth and stabilize power output from large wind farms, and hence classified as long-term investment in the transmission grid.

120 *Energy storage*

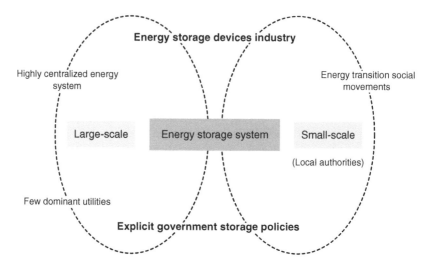

Figure 4.4 Graphic representation of the storage systems divide

right climate for energy storage. California's policy in this regard is as explicit as Germany's. However, the Californian government's primary aim is not to provide financial support to small final consumers, but to exercise soft regulation of utilities. In fact, on 17 October 2013, the California Public Utilities Commission (CPUC), a regulatory authority with all kinds of utilities within its remit, unanimously approved its proposed mandate on the Energy Storage Procurement Framework and Design Program:

> The mandate requires California's three investor-owned utilities, Pacific Gas and Electric, Southern California Edison and San Diego Gas & Electric, to collectively buy 1,325 MW of energy storage by 2020. On top of that, municipal utilities and utility districts will have to procure energy storage that equals to 1 percent of their projected 2020 peak electricity demand and do so by 2020. All the energy storage projects that are under contract under this mandate will have until the end of 2024 to complete their installations.
> (Wang 2013)

The mandate's objectives extend over many years. It does not impose a particular storage technology, and it is indifferent as to the source of energy; thus, the required amount of stored energy can be derived either from thermal or nuclear power plants. However, the priority given to load renewables in the grid is still valid. Furthermore, there are 'rules that would limit utilities from owning more than 50 percent of the total amount of energy storage to be procured across the three "grid domains" of transmission, distribution, and customer-located storage' (St. John 2013). At the same time, 'the proposal calls for utilities to consider all forms of resource ownership (utility-owned, third-party owned, customer-owned, joint ownership), including entering into contracts with customer-sited storage resources', and further states that the utilities 'may own storage assets in all three storage grid domains' (St. John 2013). This aims to stimulate initiatives for cooperation between utilities and customers sharing storage devices or services.

The last two statements serve to correct the division between small and large energy storage systems envisaged in Figure 4.4. A central role is given to the main utilities, since they are the nerve centre of the entire power system. The CPUC believes that if utilities accept behind-the-meter storage combined with the self-provision of energy, the conditions will be in place to enable great energy savings and efficiency on the demand side. During the debate on approval of the mandate, the three Californian investor-owned utilities sought to establish a threshold of higher than 50 per cent for their direct control over storage assets. But the CPUC decided to maintain the status quo, demonstrating a mode of operation by Californian public agencies linked to *institutionalism*: not imposing fixed targets, but setting thresholds within which each economic actor is allowed to find its own most convenient arrangement (Hall and Taylor 1996). This is a typical form of state/market compromise, quite far from opposed ideologies of neoliberalism or a nationally coordinated economy.[10]

Under this institutional approach, on June 2014 California's Self-Generation Incentive Program (SGIF) was re-authorized by Governor Jerry Brown. It provided $83 million per year from 2014 to 2019. This is a State Rebate Program established in 2001 that includes subsidies not only for renewables, but also for combined heat and power stations and energy storage systems, with the sole condition of being 'behind the meter'. No energy storage technology is privileged, and there is no minimum or maximum eligible system size, although the incentive payment is capped at 3 MW for a single installation. Furthermore, storage can be stand-alone or in tandem with solar or another technology.

The SGIF is considered complementary to the above-mentioned policy in helping to achieve 'the 1.3-gigawatt CPUC mandate for storage, especially since 200 megawatts of that energy storage must be customer-sited' (Wesoff 2014). Moreover, the SGIF shares the same philosophy of storage mandate: freedom of choice and a combination of technologies to include non-renewable sources of energy as well.

The German policy is slightly different: as mentioned in Section 4.1, the 'Directives concerning the support of stationary and decentralized battery storage systems to be utilized in connection with photovoltaic systems' is a policy to support final consumers through subsidies for the installation of a storage system combined with a small PV plant in a residential or commercial building.

---

10 'Differences in speed of change towards sustainable energy system between countries are related to the depth of market liberal paradigm and associated economic and political institutional differences; specifically, the UK as a liberal market economy with a strong neo-liberal policy paradigm is likely to make the transition more slowly than 'coordinated' or "managed" economies such as Germany or Denmark' (Lockwood et al. 2013: 31).

122 *Energy storage*

It is therefore much more oriented toward *integration between solar energy and storage for final consumers* (German Solar Industry Association 2013: 3). The complex and crucial relationship between consumers and the utilities is left untouched. Single utilities may voluntarily set up partnerships with their customers to provide assistance in the installation of PV plus battery systems. E.ON, one of the 'big four' German utilities, has created a special programme to help its clients use the incentive for storage. This is an initiative undertaken by one utility without any pressure from the Federal Government directive.

The German directive can be framed within traditional policies to support final consumers of green products or services like cars and domestic appliances. The subsidy also came about because of pressure from the national battery industry, which is particularly powerful in the vehicles sector. We may presume that in Germany the car industry had pressured the government to pass a measure on domestic energy storage, which can absorb the production of a very mature sector like battery manufacture, especially lead acid batteries. Rather than utilities, which do not have a common strategy, an important role has been played by the storage devices industry. And this may be a further difference with respect to the USA, where (a) utilities are more constrained, and (b) technological neutrality and a variety of storage devices prevail.

The situation can be organized into a continuum along which the four cases (plus others later) can be located temporally (see Figure 4.5).

The arrow indicates progression in the state's intervention in energy storage. It is not at all an evaluation of the best policies. In fact, most of them are recent and temporary. For example, in Germany, the directive on PV storage subsidies has a maximum annual budget, and it is experimental. The government's intention is that after two years of application, careful evaluation will be made of whether to continue with the measure (see Martinot 2015). Nevertheless, the arrow is an attempt to classify each country's policy style.

Italy and France appear to have no explicit storage policy. Of course, there is some regulation of special cases like hydroelectric plants – an energy storage form with which both countries are well equipped. The dominance of Alpine

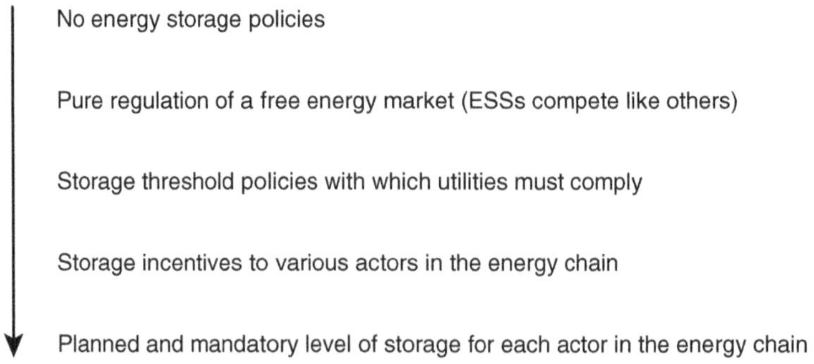

No energy storage policies

Pure regulation of a free energy market (ESSs compete like others)

Storage threshold policies with which utilities must comply

Storage incentives to various actors in the energy chain

Planned and mandatory level of storage for each actor in the energy chain

Figure 4.5 Continuum of storage policy intensity

hydro power may have been an indirect hindrance to the introduction of small, demand-side storage systems. Another more direct obstacle is probably the fact that both countries had a strong power provider operating as a monopoly. In recent years, Italy has moved more quickly than France toward unbundling the electricity sector, but the legacy of a centralized system is still very influential because of lock-in and path dependence (Carrosio 2013).

A case which approaches pure regulation (the second level down the arrow) is the state of Texas. In 2011, the state government passed a 'law (Senate Bill 943 by Carona) [that] clarifies the right of storage resources to interconnect to the grid, and to sell energy or ancillary services to the wholesale competitive market (ERCOT)':[11] 'This is a significant piece of energy storage legislation as it states that storage is given all the same interconnection rights as any other generation asset that is allowed to interconnect, obtain transmission service and participate in electricity markets.'[12] In political economy terms, this seems a case of conformity with a pure liberal model, according to which every service – even energy storage – can be a free entrepreneurial activity with the right to compete in the market.

California, on the other hand, appears less dominated by the free market creed. The Californian public showed the will to impose some external limits on the actions of utilities, and indirectly on power plant operators. This has been referred to as a compromise between state and companies. However, in this specific case the classification is not easy because the Californian government also has a tradition of providing subsidies to consumers for renewables, updated with the recent inclusion of storage devices. Evidently, California occupies two positions on the above continuum, demonstrating once again that policies can be complex and not always congruent with traditional state/market integration patterns.

Germany, although it is undoubtedly the world leader in small-scale renewables, exhibits a more limited range of action on storage policies. The balance of interests in the energy field is complicated in a country where public and private and local and global coexist: two of the 'big four' utilities (E.ON and RWE) are investor-owned companies, and two (Vattenfall and EnBW) are respectively owned by the Swedish state and two corporations operating under public law and located in Baden-Württemberg. More or less all utilities have a multinational profile, and all are shareholders in German Transmission System companies, with the exception of the TSO TransnetBW, which is a subsidiary of EnBW. All the 'big four' utilities are still owners of several power plants fed by a variety of sources, including nuclear and renewables.

The German energy sector is nevertheless a good example of Rhine capitalism (Dore 2000): it places emphasis on having a plurality of company stakeholders, as opposed to the prominence of shareholders in Anglo-American capitalism.

---

11 http://www.cesa.org/about-us/member-news/newsitem/texas-takes-major-step-toward-energy-storage (accessed 11 July 2014); for details of Texas's regulatory regime and interconnections, see Bhatnagar et al. (2013): 9.

12 http://energystoragereport.info/energy-storage-in-texas/#more-3107 (accessed 11 July 2014).

The result of this intricacy of stakes has been a timid vanguard action regarding behind-the-meter storage systems combined with PV panels. Unlike in California, utilities are not obliged to participate in the project. This confirms the divide within the power storage system between the block of incumbent utilities and the citizens' movement allied with important sectors of government and battery producers. In the end, a comparison with California can be made in terms of two features:

- the degree of separation between state agency and private companies – paradoxically, when they are more distinct, as in California, the former can more easily impose virtuous regulation on the latter; this represents the positive aspect of *ownership unbundling* (Pollitt 2007);
- the degree of alignment of economic interests with technological objectives – this is a precondition for niche innovation to progress to mass diffusion (Geels and Schot 2007); in a more coordinated system like Germany's, alignment is more difficult to achieve, but when it is established, possibly with the help of the public, it becomes durable and widespread.

These factors can be used to understand situations other than the four cases selected. Japan is the country with the highest investment in energy storage at both utility and household level. Furthermore, together with South Korea, it is a leading producer of electric vehicles using batteries, especially lithium ones. Before Germany, in 2012 (one year after the nuclear incident at Fukushima):

> the Japanese government announced a three-year programme of subsidies for renewables-related technologies. Of this, reports The Lithium Review, around 20 billion Yen, or roughly USD$212 million or UK£140 million is being spent on stationary lithium-ion battery energy storage. One of the key subsidies is to cover one-third of the cost of domestic systems and fuel adoption of energy storage at a household level.[13]

This covers only the announcement; implementation began in March 2014, when:

> the Japanese Ministry of Economy, Trade and Industry (METI) announced that it would begin accepting applications for a generous subsidy that it is awarding to consumers who install lithium-ion battery storage systems along with renewable energy systems on their homes or businesses. The subsidy will spark nearly 100 MW of energy storage capacity to be installed in 2014, according to an energy analyst with IHS. METI is subsidizing up to two-thirds of the upfront costs of 1-kWh or larger stationery lithium ion battery systems that consumers install to back up their PV capacity. The maximum

---

13 'Japan's big investment in energy storage', *Energy Storage Report*, March 30, 2013; http://energystoragereport.info/japans-big-investment-in-energy-storage/ (accessed 4 July 2014).

subsidy is set at ¥1 Million (about US $10,000) for consumers and ¥100 Million (about US $1,000,000) for commercial organizations. In all, METI has set aside about ¥10 Billion (about US $98 million) for the program.

(Runyon 2014)

The parallel with Germany is evident. The design of the energy storage policy is similar. Incidentally, Japan is classified by Dore (2000) as within the Rhine model of capitalism. The electricity system is not dissimilar: in Japan, the sector is dominated by ten privately owned utilities which vertically control almost the entire power chain in the region, where they have a monopoly on transmission and distribution (Navarro 1996). Most generation plants are in the hands of small wholesale companies whose owners are municipalities. Thus, the PV plus small storage station combination would harm the Japanese utilities less than German ones that, conversely, own most of the power plants.

Furthermore, Japan, like – or probably more than – Germany is a case of good alignment among research and development, market growth and government targets (Watanabe, Wakabayashi and Miyazawa 2000). This alignment has been propelled not only by the general issue of national security since the Fukushima incident, but also by the proverbial common understanding between METI and Japanese companies – in this case, those involved in the production of storage devices (Tomita 2014). This situation is similar to the one observed for the Japanese PV panels industry:

> A special feature of the institutional structure of production of the Japanese PV industry is that a few large and integrated companies bundle the whole or at least large portions of the PV value chain inside their own company. These may include solar cell, module (many solar cells combined together), BOS (balance-of-system) components such as inverters, power electronics and sometimes even the installation and maintenance of the PV systems are offered from the same company.
>
> (Shum and Watanabe 2007: 1192).

If we add lithium battery technology to the BOS components, the picture presents many similarities with the history of PV panels. The above quotation implicitly indicates a difference between Japan and the USA in the diffusion of on-grid PV panels which may also apply to the energy storage industry. In fact, Shum and Watanabe (2007) assume that behind-the-meter PV panels have penetrated the Japanese market to a greater extent, not only because of better alignment, but also because the PV package has been developed as a *manufactured technology* – a standard product with a low level of customization, and therefore ready for mass diffusion. The opposite, according to Shum and Watanabe (2007: 1,194), is *information technology*, which has 'numerous applications supporting different business models' that are supposedly slower in diffusion.

Japan and Germany provide subsidies for a unique combination of PV panels and batteries. By contrast, California – reflecting a more *technologically neutral* and less coordinated economy – leaves the option of whether or not to combine a storage device with PV open, and it does not privilege any storage technology. This picture is also confirmed for public action on utilities: Japan, like Germany, does not impose any obligation on utilities, while California imposes a specified power storage target on each of the three main utilities operating in the state. The difference is remarkable, but it is difficult to say whether these are opposite models – the Nippon-Rhenan versus Anglo-American types of capitalism.

The situation of Japan with respect to California and Germany helps to clarify the current absence of policy in Italy and France: the two countries are well equipped with hydro power, as a source and form of storage, aligned with a centralized power distribution system. Moreover, though to different extents, they have a single incumbent utility. This combination leads to caution toward demand-side power storage. Both countries lack a national 'champion' battery industry. There is indeed FIAMM in Italy, but it is an isolated, 'pocket-size' multinational without the common understanding with government typical of Rhine capitalist countries.

4.5 Thermal energy, the Cinderella of storage

Thermal energy storage (TES) can be defined as: 'a technology that stocks thermal energy by heating or cooling a storage medium so that the stored energy can be used at a later time for heating and cooling applications and power generation' (IEA-ETSAP and IRENA 2013: 1) and divided into the following types:

> *Sensible heat storage*, in which the temperature of the storage material varies with the amount of energy stored, and *latent heat storage*, which makes use of the energy stored when a substance changes from one phase to another by melting (as from ice to water).
> (Hasnain 1998: 1,127; emphasis added)

We should add a third type: *thermo-chemical storage* (TCS), which uses chemical reactions to store and release thermal energy. The second and third types are still at the experimental stage: 'At present, TES systems based on sensible heat are commercially available while TCS and Phase change materials (PCMs)-based storage systems are mostly under development and demonstration' (IEA-ETSAP and IRENA 2013: 3).

Systematic exploitation of thermal energy storage can be divided into four classes:

1   Energy uses – Energy storage is used instrumentally in an industrial process whose final purpose is the generation of electricity. It is a complement to variable price/production power, as happens for pumped hydroelectric storage. In this case, a heat pump converts excess power into heat to be stored and used later for conversion into electricity; a variant of this class is molten salt thermal storage, also known as *concentrated solar thermal power*: heat storage is achieved by means of a set of mirrors positioned so as to reflect the

sun's rays onto a heater containing a molten salt; the heat thus concentrated feeds a turbine for the production of power. Curiously, this is an old technology, developed in the nineteenth century, but soon abandoned because it was impossible to compete with coal at the time (Ragheb 2014).

2   Conditioning uses – Thermal storage is located underground in natural cavities, boreholes, or in contact with an aquifer. Equipped with a heat pump, it increases the difference in temperature between underground and ground level.[14] This system can also be used to cool buildings during warm seasons. Moreover, it can be connected with industrial uses or district heating/cooling systems. A special variant is *geothermal energy*, whose storage relies on the existence of natural hot water or rocks at variable depths.

3   Domestic uses – Thermal energy storage is a small-scale process involving boilers placed inside buildings to provide constant hot water for sanitation and washing. Insulated tanks are used to store water heated in a number of ways: by electric immersion heaters, heaters fuelled by gas or biomass, or rooftop solar collectors. Combining hot water sources is possible, but it is rarely used when there is a grid connection (district heating systems). In some modern buildings, sanitation hot water is inserted into a dual system for cooling and heating the entire building using heat pumps and installations to capture underground latent heat (*enthalphy*). This is a variant of the system illustrated in point 2 above. Thus, these systems may be single-purpose (only heating), dual-purpose (cooling and heating), in-grid or off-grid, and may or may not be able to exchange cold and heat between buildings with different functions.

4   Construction sector uses – A further thermal energy storage system concerns the insulation of buildings. This is usually considered separately from other forms of thermal energy storage because it is based on techniques that are quite different from those employed in the production and conservation of cold and heat: one focuses on heat pumps and exchangers, the other on the performance of insulation materials. The former are much more engineered, while the latter are based more on combinations of different technologies. It is also easy to envisage applications in different product sectors, such as mechanical engineering and building construction.

These classes of thermal storage use can be summed up in a more stylized manner as in Table 4.4). The difference compared to power storage is that the techniques are simpler, at least for sensible heat storage. They simply take the form of, for instance, glass demijohns stored in the roof space to heat and keep water warm on a daily

---

14 'The idea of using a heat pump to extract heat from the ground was patented by the Swiss Heinrich Zoelly in 1912 [. . .], and was being used (with rivers and groundwater from wells as a heat source) by the 1930s in America and Switzerland. The "closed-loop" heat pump, where a carrier fluid or refrigerant is circulated through a loop of pipe in the subsurface to extract heat from rocks and sediments, was first constructed in 1945 by Robert C. Webber in Indianapolis, USA' (Midttømme et al. 2008: 94).

basis during the hot season – a system very common in houses in Mediterranean rural areas. The most sophisticated plants are large underground natural or artificial containers equipped with a heat pump. The drawbacks of sensible heat storage are its low energy density (large volumes achieve only low differences in temperature) and the fact that the storage process is not *isothermal* – it does not allow regular release of the heat or cold. The other two methods of thermal energy storage, PCMs and TCS, do not have these drawbacks (Sharma et al. 2009), but they are more expensive and much more technically complex, and therefore less accessible to common people or small organizations. That is an important aspect to bear in mind when considering smart thermal energy grids.

It should be noted that the difference between sensible heat storage and the other two systems concerns control of the thermal process: the two more sophisticated methods allow better planning of timing and temperature. They are industrial processes in the full sense of the term: reproducible on a large scale, energy-intensive, needing a protected environment like a factory, and relatively easy to transport. In this sense, they can be reproduced everywhere, while sensible thermal storage relies on an immediate source of heat and specific geological conditions, at least for large underground systems.

The most organized source of information for thermal energy storage is the US Department of Energy Global Energy Storage Database.[15] This open-access website is continuously updated, so data will change over time. Nevertheless, the general picture is clear: in August 2014, there were 1,000 energy storage projects, 766 of which were operational. On the same date, 162 projects were classified as thermal storage, reducing to 105 if restricted to operational and verified cases. The statistics were as follows: seventy-five ice thermal storage projects, seventeen chilled water storage projects, eleven heat thermal storage projects and six

Table 4.4 Forms of thermal energy storage according to their sources, methods and uses

Sources	Storage methods	Uses
Solar collectors Power plants Waste heat	Special containers in industrial complexes	Concentrated solar power plants Heat pumps District heating
Underground stable temperature settings	Boreholes (B-utes) Aquifers (A-utes) Caverns	District cooling/heating Cooling or heating buildings Support to industrial processes
Solar collectors Biomass Fossil fuels	Boilers/hot water tanks	Daytime and nighttime hot sanitation water
Natural and artificial materials	Materials' properties and construction techniques	Improving insulation/quality of dwellings

15 See http://www.energystorageexchange.org/projects (accessed 5 February 2016).

molten salt projects. The majority of projects concerned storage of cold in the form of ice or chilled water; they were located in the USA and had two purposes: to accumulate thermal energy during off-peak times, and to maintain a reserve to cope with extreme situations like blackouts or high demand for air conditioning. The database category 'heat thermal storage' comprises heterogeneous projects: most of them are concentrated solar power plants, but three are *electric thermal storage systems*, which 'consist of high-density ceramic bricks heated to a high temperature with electricity, and well insulated to release heat over a number of hours' (Buchholz and Styczynski 2014: 54). These systems are small interconnected boxes installed in houses: the innovative feature of the projects included in the database is that there are:

> electronic controls for these heating systems that allow individual heaters or electric boilers to be switched on or off rapidly (within seconds) of the receipt of a control signal by VCharge's Network Operations Center from the area's grid operator, PJM Interconnect.[16]

The last and least numerous category is easily identified: it includes all concentrated solar power plants used for the accumulation of heat directed by mirrors to a receptacle containing molten salts, usually sodium and potassium nitrate. There are only a few of them, but if we look at the total of the projects included in the database but classified as unverified and non-operational, the number of cases triples to twenty-two – a rate of 'provisional' much higher than that of the other thermal energy storage technologies. This testifies that molten salts comprise the most dynamic and laborious socio-technical package. According to the US database, most of these plants are in Andalusia in southern Spain. This is currently the only major novel initiative in thermal storage technology after the widespread diffusion of ice and chilled water storage in the USA.

Very small thermal storage systems are not included in the US database. Thermal storage, as mentioned earlier, is very old and widespread in all industrialized countries in the form of household boilers. Moreover, in the US projects collection, the number of relatively large underground thermal storage systems is probably underestimated. They are most common in northern Europe and concern two types of storage: systems using natural underground temperatures – geothermal energy storage – and heat storage combined with solar collector plants, sometimes integrated with other sources of energy like biomass or fossil fuels.

The first type, according to a report issued some years ago (Hendriks, Snijders and Boid 2008), is most widespread in Sweden, followed by Holland, Germany and Belgium. According to the same report, thousands of small plants

---

16 The quotation is taken from the Pennsylvania ATLAS (Aggregated Transactive Load Asset) case classified in the US DoE database. Vcharge is a company that sells both heater systems and electricity; PJM Interconnection is a regional transmission organization that coordinates the movement of wholesale electricity in the whole or parts of several US states.

serving hospitals or commercial buildings can be found in each country. More recent calculations estimate that there are about one million plants in Europe (Meneghello 2012). Southern European countries are almost absent from these lists, for reasons that are not immediately apparent. Geological explanations do not seem plausible: such technology is dangerous in areas where radon gas is present, which is not typical of southern European countries. Subsidence or underground water pollution is a risk for heat pumps using aquifers situated on low land at all latitudes: Holland as well as the Po Valley in Italy (Bonte et al. 2011). The marked differentiation is probably linked to the traditional technologies used to heat buildings (*past dependency*). In northern Europe, the urgency of finding ways to heat houses has given rise to a wider range of solutions, not least low-enthalpy geothermal systems. These divergent technological paths may be explained by legal considerations (simple authorization procedures) and sometimes incentives for thermal storage, which have been lacking or belated in countries like Italy.[17] Furthermore, the cost of installing a geothermal plant in an existing building is still very high, both for drilling the borehole and for converting the heating system, since radiators must be replaced by fan coil units or underfloor heating/cooling. Such indoor climate control systems are much more convenient for new or fully renovated buildings.

The second type of thermal accumulation system which appears neglected in the US database is heated water storage connected to solar collector plants. In this case, there are two ways to store heat:

- central solar district heating plants – solar collectors deliver heat to a central storage device; with large seasonal heat stores, such plants can cope with more than 50 per cent of the total heat demand;
- distributed solar district heating plants – solar collectors are placed at suitable locations and connected directly to the district heating primary circuit on-site; these plants rely on the district heating network for storage (Pauschinger 2011: 5).

Theoretically, there is a large difference between these two methods in terms of storage: in the former case, there is a large heat store able to retain warm water from one season to the next; in the latter, the network of pipes works as a store, but obviously not from season to season. A distributed solar district which directs the heat from many rooftop collectors to a central heating plant does not involve long-term storage. Furthermore, it does not avoid the need to construct a centralized heating plant because the amount of heat provided by the sun collectors is insufficient. What it does avoid is the need for a large expanse of ground-mounted collectors, which are typical of centralized heating districts.

---

17 In Italy, the idea of introducing special incentives for thermal energy, which could have helped thermal energy storage projects such as geothermal, came too late, after the campaign against renewables had already provoked a dramatic halt on all subsidies (see Meneghello 2012).

In practice, there are mixed cases. In Germany as in Sweden, district heating systems are fed both by ground- and roof-mounted solar collectors, allowing the two types to be integrated. In the Swedish town of Vislanda, a net metering contract has been developed between the district heating provider and single residential units with solar collectors on their roofs (http://www.solar-district-heating.eu). This method of measuring the self-production of energy is absolutely identical to that for units equipped with PV panels, called *net metering*. A small substation at user level must be installed to measure the flow and temperature of the water. It is bulkier and more expensive than a power meter: in fact, the applications are more common in multi-family or municipal buildings, which can more quickly amortize the cost of the meter. Nevertheless, the advantages are clear, in both economic and environmental terms: residential units benefit from a reduced district heating tariff, and there is an incentive to reduce hot water consumption because usage is measured more accurately.[18]

The optimal solution would involve not only a large container of hot water to be used on a seasonal scale and coupled with district heating, but also small storage devices plus solar panels in single buildings connected to the grid. This will maximize heat savings, reducing the need to feed the district heating system with fossil fuels, not to mention biomass, which leads to a high environmental impact in terms of CO_2 and particulate emissions. The combination of a large storage system plus the net metering of rooftop solar collectors plus domestic storage arrangements creates the conditions for a *thermal energy smart grid*. The central heat plant thus works only in a *subsidiarity regime*. This is not easy to accept, especially in those district heating systems where large recent investments have been made to increase the efficiency and capacity of a centralized plant fuelled with methane or other gases. Not only are the initial investments to be reimbursed to the banks huge, but also the gas purchasing contracts are rigid: 'Long-term natural gas purchase contracts containing take-or-pay clauses created huge tumult in the industry' (Petrash 2006: 546).

These are probably the reasons why solar collectors integrated with district heating systems are absent in a country like Italy, where these arrangements have a good but separate path of development (see Verda and Colella 2011). In Italy, according to data issued by the IEA (OECD/IEA 2014: 113), the provision of energy using solar water heating collectors was 2.4 thermal gigawatts (GWth) – a modest contribution representing only 0.8 per cent of world provision, but nonetheless it puts Italy among the top twelve countries. Therefore, the solar heating sector is present in Italy not only at an experimental level. At the same time, according to the Associazione Italiana Riscaldamento Urbano

---

18 Because net meters provide more precise feedback about consumption, they stimulate the search for further information on energy sources, competition with others to consume less, and the use of appliances whose energy sources are more abundant or less expensive (see Abrahamse et al. 2005; Fischer 2008; Faruqui, Sergici, and Sharif 2010).

(AIRU) there are 109 district heating systems in Italy, providing 279 million m^3 of heat (AIRU 2013: 23). None of these systems use solar collectors. The first one, which is quite small, is under construction in the town of Varese. District heating systems in Italy have a strong tradition of relying on fossil fuels or biomass – about sixty rural biomass district heating systems can be found in the South Tyrol alone.

These separate trajectories – some hybrid solutions exist, but only those combining waste heat and gas, and never integrated with solar collectors – warrant explanation. One reason concerns the urban/rural dimension. Historically, district heating systems in Italy have been installed in urban areas – as indicated by the name of the association of district heating management organizations, Associazione Italiana Riscaldamento *Urbano*. In urban areas, there not many surfaces are available for mounting solar collectors, which have more rigid requirements in terms of their angle to the sun than PV panels, especially if the building density is high. In Denmark, the best-known integrated plant has been constructed in Marstal, a small town of 2,400 inhabitants on the island of Aeroe. The water of the district system is warmed by '33,000 m^2 solar heat panels on 10 hectares of land [. . . that] produce more than 50% of the heat consumption and the rest will come from biomass energy' (Andersen, Bødker and Jensen 2013: 3,352). The scale of this installation testifies to the availability of a wide-open green area near the plant and the houses. In urban areas, such large unutilized spaces are difficult to find, and the roofs of buildings have limited space. Also, the storage structure in Marstal, as in other Danish cases, takes up even more room, since caverns or other underground containers are unavailable.

A further explanation for the separation concerns the technological trajectories – some are more compatible with renewables than others:

> Fossil based Combined Heat and Power (CHP) dominates electricity generation and the heat supply in urban areas in Denmark. The recent strong development in wind power in Denmark has created a situation where in periods with good wind conditions it is less feasible to operate CHP and more feasible to operate [electrical] boilers to supply the required district heat. Relatively high district heat costs and a strong local solar collector industry caused the idea to implement large solar heating plants in connection to existing or new short-term storages in CHP plants.
>
> (Pauschinger 2011: 6)

The widespread use of wind turbines together with a vibrant solar collectors industry have made combined heat and power plants less appealing. At this point, the issue of coexistence arises: are plants fired with fossil fuels or biomass well matched with solar collectors, or is these renewables destined to completely replace CHP? The answer is hypothetical, and divides into two parts. First, it is very difficult to think, even in the long term, of replacing centralized plants relying on combustion with solar collectors. The Danish government has established an ambitious programme to provide 10 per cent of heating demand with solar collectors by 2030.[19] Therefore, integration with more traditional sources of heating will remain necessary for a long time. The main problem is likely to be how to convert the CHP plants so that they work in a subsidiary manner, complementing the privileged renewable sources.

One solution, and this is the second part of the answer, is storage, which is one way to create stronger integration between renewables and non-renewables. Storage makes it possible to avoid wasting heat from CHP plants working over long, regular temporal terms. Not only can the waste heat be stored, but also when electricity registers disadvantageous prices, it can be converted into heat and stored. *Combined heat and power storage systems* should be designed and built along the same lines as CHP plants. There is a broad consensus in the specialized literature about the feasibility of heat storage in CHP plants (Nuytten et al. 2013; Mago, Luck and Knizley 2014; Christidis et al. 2012). Nonetheless, CHP storage systems currently remain at the experimental level.[20] The crucial issue seems to be the creation of high-temperature storage structures, which make it easier to convert or re-convert heat into power. In some ways, this is what happens with concentrated solar power (Denholm and Mehos 2011). However, concentrated solar power's use of high temperatures leads to centralized, supply-driven energy systems – an approach very different from heat storage with small-scale solar collectors.

This digression on the separation between solar collectors and district heating systems, which is emblematically represented in Italy, leads to an already-mentioned distinction, that between high and low *technologically dense systems*. If *low* heat storage is a way to combine old and new sources of energy, *high* heat storage, intended to give the greatest flexibility to power systems, again exacerbates the dichotomies between low-density and high-density technologies and large versus small plant scales. Thus, we return to heat's ancillary function with regard to power, although 'heat and chemical storage may have greater capacities than electricity storage' (Barrett and Spataru 2013: 679). The market growth in modern renewable energy technologies for heating and cooling 'continues to lag behind the power sector' (OECD/IEA 2014: 28).

The subordination of heat to power is not justified by technological arguments, and calls for intervention by policies and consumers. In terms of policy intervention, the matter is straightforward: coupling a local tradition of district heating

---

19 'Solar district heating growing rapidly in Denmark', *District Energy*, 7 October, http://www.districtenergy.org/blog/2013/10/07/solar-district-heating-growing-rapidly-in-denmark/ (accessed 14 April 2015).

20 'The "FleGs" research project (Increasing the flexibility of combined cycle power plants by using high-temperature thermal storage systems), funded by the German Federal Ministry of Economics and Technology, is developing a plant scheme with a novel solids-based thermal storage facility. The aim is to increase the deployment flexibility of Combined Cycle Gas Turbine/Combined Heat Power plants to compensate for the fluctuating input of renewables-based electricity to the grid, to stabilise the network and to make balancing energy available'; http://www.rwe.com/web/cms/en/1265036/rwe/innovation/projects-technologies/power-generation/fossil-fired-power-plants/combined-heat-and-power/, accessed on 24 April 2015.

with a deliberate national government choice on energy transition has led to the adoption of suitable thermal storage initiatives. The case of Denmark is paradigmatic: it 'is increasing the reliability of its energy supply by combining variable renewable electricity with CHP and district heating, and has made this practice a cornerstone of its energy policy' (OECD/IEA 2014: 28).

The role of heat storage consumers is more difficult. In the literature, largely framed in technological and economic terms, consumers rarely have a pivotal role. Most of the time they are passive recipients of schemes devised by institutional actors. Net metering for district heating systems certainly magnifies the deliberate actions of single users, inducing a more rational attitude to energy. Rural communities in central Europe, from Sweden to the Alps, have organized integrated systems for district heating. The pivotal role of municipalities or local cooperatives is decisive in such projects. The relationship between the virtuous institution and the final user is not always balanced; the absence of integration of household heat storage with district heating systems is indicative of the dominance of cooperative bodies over single consumers. That is not necessarily a bad thing, especially when the environmental impact of these collective schemes is lower than that of single-dwelling boilers.

Nevertheless, the actor side is fundamental for the capacity both to use the new technologies and to reduce the levels of heat/cold consumption. With regard to the former point, effective progress can be made by deploying smart controls and digital devices (Brandon and Lewis 1999). These allow strict self-monitoring of consumption and comparison with other households accepting reciprocal control. With regard to the latter point, the expectations are very low: people are reluctant to reduce their levels of energy consumption (Abrahamse et al. 2005). Hopes of reductions in the growth of energy consumption rely on the actions of energy-obsessed minorities (Osti 2006) or of groups committed to social marketing (Haq, Cambridge and Owen 2013); both have developed a systematic range of virtuous actions. For food, this is easier, because the leeway is very wide (see Chapter 2). For energy and for people living in urban areas, the room for manoeuvre is much more limited because there is a shortage of space in which to install thermal energy storage devices or solar collectors. Nevertheless, there is a possibility for a large number of people to make small gestures: every household or building has a boiler that can be fuelled in various ways.

Realistically, integration with targeted policies is the best way to overcome space limits and the laziness of consumers: 'Where a carbon charge exists, heat users tend to seek low-carbon fuels. Consumers in Denmark, Japan, and the United Kingdom can choose green heat via voluntary purchasing programmes, but options are relatively limited compared to green power purchasing' (OECD/IEA 2014: 28).

Hence, the solution is to increase the options for heat/cold storage compared to those available for green power, considering that heat storage options are less manageable: they require dedicating often scarce space for containers and solar panels, they are less transportable, and finally, less versatile. Faced with these difficulties, most governments are inactive in giving incentives for thermal storage and negotiating with the consolidated interests of utilities, especially those organized as fossil-fuelled district heating systems. Indeed, district heating systems have a lot of potential, enabling strong socio-technical integration between plant and the grid, and between a centralized source of heat and a network of consumers. Modular storage devices can be inserted at all levels of district heating systems. The technical means – net meters and in-home displays – are available, but they have yet to be deployed on any great scale by utility boards.

4.6 Conclusions

The conclusions to this chapter are based on the same questions addressed for food storage. The first question is whether energy storage is a category able to highlight new or neglected social phenomena. The answer is probably disappointing: at present, innovations are almost entirely confined to the batteries industry, which is pressing institutions for the mass introduction of storage devices. Consumers are not exhibiting the same urgency nor are they joining together to make collective purchases of storage devices as they did for PV panels. They are rather impassive to marketing by utilities and industries. Although community energy storage systems are appreciated by researchers (Parra et al. 2015), they are rare, experimental and have uncertain results (Tweed 2013). Some countries are more dynamic in combining their own manufacturing traditions with growing environmental concern among the public.

The main innovation probably involves the modular reorganization of life; energy storage can be integrated at so many levels and in so many forms that human existence can be rearranged to increase autonomy and personal satisfaction. The symbol of this modular change is the automobile. There are prospects of increasing the autonomy of vehicles by means of a new generation of accumulators. These are currently mainly based on improvements in lithium-ion batteries, but many other technological packages are earning attention. The same cannot be said of thermal storage. In this case, the analysis has shown that the main social arrangement – the district heating system – is resistant to changes induced by advances in the storage of heat. The desirable combination of a grid supply and self-storage has not yet been achieved even in the most advanced case of Denmark. There are only a couple of net metering solar district heating systems, in contrast to the large-scale investment in thermal utilities.

Nevertheless, the appeal of modular systems provides a way to answer the second question: whether 'storage has been an ideal type with heuristic capacities'. Promising developments in modularity in the case of automobiles, for instance, or the lack of them in the case of district heating system, for example, are also a *social pattern* (Gigerenzer 1997) and a measure of *social networking* (Newman 2006). Traditional ways to interpret energy policy and provision are based on binary oppositions – fossil versus renewable energy sources being the most evident. Modularity is exactly the contrary of a binary model because it is based on *gradualism* (Urry 2010). The incoherence of some governments in promoting energy storage disappears when a modular model is applied. The governments most active in energy storage are those able to act at different levels with different instruments. As we have seen, Japan and Germany have privileged behind-the-meter storage, whereas California has set boundaries on the actions of utilities without abandoning subsidies to final consumers. California, in fact, is the most innovative case in terms of adopting energy storage on a modular pattern through the use of multi-level and multitasking measures, technological neutrality and state/market compromises.

The third question is whether energy storage is an original way to respond to the ecological crisis, a *third way* between merely increasing efficiency or pursuing drastic self-reduction in consumption. Energy storage is one way to keep the remaining quantities of oil underground, to limit CO_2 emissions, and to save a great amount of money for energy final users. Most of these technologies are still experimental, leading to caution among investors. But some energy storage tools are so evidently useful – such as heat storage – that the social and institutional tardiness is surprising. This is probably because fossil sources of energy are still so cheap that storage takes the form of a reserve field: a suitable technological solution as a hedge against remote adverse periods. This makes the drive for energy storage innovations difficult to appreciate now. Their adoption demands more courage; the storage devices industry does not lack this, but consumers and utilities are still waiting for external action, presumably in the form of subsidies, compulsory measures or a traumatic event like the collapse of the fossil resources system.

5 Long-term life storage

So far, we have considered forms of storage with a limited time span. Food, water and energy storage were developed to conserve resources from day to day or season to season – or even for a number of years, according to the biblical story of Joseph, which recounts that seven years of famine would be followed by seven years of plenty. Seven years is a very long period for storage of a valued item, whose immediate consumption is sacrificed with a view to its better use in the future. But are there goods that can be stored with a view to a benefit that is much more distant – practically for ever?

Life forms fall within this special long-term storage category. In many discourses on environmental protection, frequent reference is made to future generations. This means, according to a particular standpoint, the conservation of life – human life in this case – but it can be extended to the continuity of animal life or habitats as well. When we talk of future generations or endangered animal species, the time horizon becomes undefined, almost infinite. What needs to last is not a single person or body, but complex sets of relationships among species and their environment – *ecosystems*.

Some applications of short-term storage have a precise instrumental objective: to set aside food today so that it can be consumed tomorrow. The time scale can vary greatly: for example, an artificial water basin may become useful during a flood occurring in thirty years' time. In this case, a calculation is made, usually accompanied by a cost forecast. Apart from these various forms of storage whose future utility can be calculated, there are nature conservation activities whose time span is difficult to establish. They pertain to the sphere of *life storage*: activities that contemplate conservation for an indefinite period of genetic material that can be converted into a living organism, known as the *germplasm*:

> a collection of genetic resources for an organism. For plants, the germplasm may be stored as a seed collection, sometimes stored cryogenically or in a nursery or botanic garden. For animals germplasm may be cryogenically stored as semen, eggs or fertilized early-stage embryos.
>
> (Steffen et al. 2009: 188)

138 Long-term life storage

After a very brief definition, Steffen et al. move rapidly on to the issue of conservation, treating different types of germplasm storage as if they were an integral aspect of the topic.

We will adopt a threefold typology to classify types of life storage. One class, the gene bank, has already been covered; we will add two more: botanical gardens and nature reserves. They can be arranged into the continuum shown in Figure 5.1.

Their location along the continuum depends on the degree of isolation from the context of the storage structure. A gene bank is an entirely artificial place where germplasm is kept under controlled moisture and temperature conditions. It is therefore enclosed within walls, and may even be located underground in order to improve the prospects of conservation and to limit costs. For botanical gardens and animal sanctuaries, the degree of isolation is certainly less. These are open-air places, sometimes greenhouses or aviaries. In these cases, contact with the external environment is much more prevalent, including public visitors. Finally, there are nature reserves. In this case, we need to focus not on the precise term used for them in each country, such as park, protected area, reserve or biosphere, but on the basic intention: a form of open-air storage where animals and plants (and regrettably, in the past, human beings as well) are accorded special respect within a bounded area. The rules for managing and monitoring such protected areas vary greatly, but the principle is a more or less permeable boundary that separates the reserve from the rest of the environment. What happens within that border is less regulated than in the other two forms of storage, relying more on spontaneous natural processes.

This chapter is divided into four sections. Section 5.1 deals with the forms of biodiversity storage presented in the continuum in Figure 5.1. Section 5.2 considers the interface between nature and agriculture according to two models: land sparing and land sharing. Section 5.3 discusses whether and how storage for typical productive purposes can increase biodiversity for broader and longer-term aims. The literature on these three processes is immense and very technical, and it is not our task to try to cover it all in this chapter. Instead, we will outline the socio-institutional frames of biodiversity conservation so that the conclusions in Section 5.4 can apply the three recurrent issues discussed in this book to life storage: (1) whether it can highlight new or neglected social phenomena; (2) whether it is an ideal type with heuristic capacities, and (3) whether it is intermediate between the shallow and deep views of environmental issues.

Figure 5.1 Classification of biodiversity storage forms according to the in situ–ex situ continuum

5.1 Nature reserves, botanical gardens and seed banks

Biodiversity protection can be considered as one the three main components of the environmental issue, the other two being the pollution of elements vital to life forms, including humans (water, air and soil) and the irreversible depletion of natural resources. Biodiversity is the most elusive factor because its usefulness for humankind is less directly evident. Pollution is the diffusion of substances in the environment in quantities and in compositions that threaten environmental health. Typical examples include the presence of pesticides in fruit and vegetables, bacteria in potable water and particulate matter in the air we breathe. Depletion is the exhaustion of basic resources like fertile land or fresh water, or even a survivable atmosphere.

Biodiversity is also more elusive because it is an abstract concept linked to the taxonomy of natural science: it refers to the number of species and the variety within single species present in an ecosystem. The variety of ecosystems in a landscape is also a source of biodiversity. But it is evident that measurement of the number of species, their variability and the scales to apply are determined by disciplinary standards (Turnhout and Boonman-Berson 2011).

Nevertheless, biodiversity protection, under the rubric of safeguarding natural sites, was one of the primary objectives of the modern environmental movement. At the time of its greatest expansion, the 1980s, two main tendencies were identifiable: the socio-political wing, which was more concerned with environmental injustice in urban areas, and the conservationist wing, which was more committed to safeguarding charismatic endangered species such as wolves, bears and eagles (Diani 1995). The panda was a symbol of this trend, and appears in the logo of the World Wide Fund for Nature (WWF), the group most representative of the conservationist wing. This wing's focus can be seen as having successfully spurred widespread concern about the loss of species.[1] However, biodiversity is more abstract and therefore less visible than other types of environmental concern because it does not target single species, but entire ecosystems. It therefore claims to be more scientific and less emotional than the old endangered species protection movement. This is probably one of the reasons why it has been confined to narrow circles of experts and aficionados. There is therefore a disjuncture between the cold abstractness of biodiversity as a concept promoted by an élite of experts, and the warm appeal of natural sites and wild animals able to excite international public opinion. International associations like the WWF have millions of members, and other Western nature protection organizations also have strong popular followings (Van Koppen and Markham 2007). These other national organizations are more focused on natural monuments (heritage) and the enjoyment of nature and rural sites through activities like hiking, picnicking and excursions

---

1 This success is testified by the growth in initiatives to protect threatened species like the panda even in authoritarian countries, by the spread of organizations like the WWF, and by the measures adopted by most countries to safeguard such animals in one way or another (Van Koppen 2006).

(Van Koppen 2006). They certainly have conservationist attitudes, but these find expression more in sentiments related to national identity and opportunities for pleasant open-air activities.

These brief remarks on the environmental movement show that nature conservation is not separate from the sentiments and practices of a large proportion of the public. Consequently, its most abstract version – biodiversity storage – is not simply an academic categorization; all nature study activities and conservation measures are practised within specific cultural frames, and as a result are more or less familiar to the people within them. In turn, this familiarity is based on learning and frequency of contact: 'The basic assumptions of the tradition are acquired less from formal principles than from familiarity with its historical exemplars' (Barbour 1974: 186). So to what extent are nature spots historical exemplars – that is, sites that are able to recall experienced knowledge, similarity of feelings and common values? In the above continuum of wildlife storage forms, those are probably more familiar lie in the middle – botanical garden/animal sanctuary – compared to natural parks on the one hand, and seed banks on the other.

This statement is backed by two arguments linked to familiarity: one concerns tradition, the other the domestication of nature. Undoubtedly, botanical gardens are much more ancient than the other two forms of storage. They were first created many centuries ago, initially as enclosures associated with royal or noble palaces, then as living classifications of nature's objects, assuming an important role in universities. They have practical purposes as well, especially for the production of medical herbs. Thus, aesthetic, scientific and practical motivations combine in the creation of these gardens. They are usually walled in order to protect them from intrusion by animals or people. Stockades, irrigation systems and specially formulated soils provide more favourable conditions for plant reproduction as well, especially where their purpose is to collect foreign species for which the local climate and soil are not optimal.

A variant of the botanical garden is the arboretum: a collection of exotic ligneous plants whose history is even longer than that of botanical gardens because of the monumental significance attached to certain large trees. In these very old institutions, admiration and fear of nature are balanced. Evidently, 'storage' is not the most appropriate term to describe the functions of these ancient institutions, because it is too closely tied to the instrumental aims of the accumulation of goods, whereas gardens are more often places of enjoyment – the pleasure of contemplating a landscape and spending time among living beings. These sensations are very far from those experienced in a storehouse. In conclusion, the sense of familiarity is balanced by sensations of surprise and concern for the natural world. Sentiments toward nature are a mix of fear and admiration, well represented by the word 'awe' (Gregoire 2014).

The accolade of the most ancient botanical garden in the world is disputed: 'The origin of modern botanical gardens can be traced to European medieval medicinal gardens known as physic gardens, the first of these being founded during the Italian Renaissance in the 16th century' (Borokini 2013: 88). Later, they collected the fruits of world exploration and colonisation, becoming 'imperial engines of appropriation' (Forbes 2008), and were located in those countries that were more zealous in those activities. Today, according to the database of Botanic Gardens Conservation International (BGCI),[2] there are more than 3,200 botanical gardens throughout the world. Many are small and not always well-managed. The more active and robust ones number around 500, if we consider those that have signed the International Agenda, 'a policy framework for botanic gardens worldwide to contribute to biodiversity conservation, particularly as it relates to the implementation of the Convention on Biological Diversity'.[3]

The geography of botanical gardens shows that such institutions are present all around the world; one or two gardens can be found even in small and poor countries. The English-speaking countries account for about 1,200 cases, then there is a block of mainly western European countries (Germany, France and Italy) with around 100 gardens each. Large countries like China, Russia and India have slightly more than the western European ones, but as a proportion of their surface areas, this is much less than small Continental countries like Holland, Belgium, Austria and Switzerland, which have more than twenty botanical gardens each. In the Far East, particular mention should be made of Japan with sixty-five botanical gardens and South Korea with fifty-four.

These data have been taken from the same source, the BGCI database. It is possible that it has a slight bias in favour of Western countries, especially English-speaking ones. But in general, the geographical distribution of botanical gardens reflects national traditions: they usually were the result of the political will of a university, a lord or a scholar. Furthermore, botanical gardens are urban structures. The ancient herbariums of monasteries should be excluded from this, since most of them were located in the countryside. However, European urbanization is very extensive, and some towns arose around a monastery. In other words, a botanical garden's origin is always linked to a *civilization point* – by which is meant a place where nature has been shaped for a variety of human purposes.

This brief survey of the history of botanical gardens as the original stores of a domesticated nature are useful for understanding the other more recent type of storage: nature parks. These can be evaluated according to their distance from the prototype of the garden. They are distinct from it for three reasons: (1) they are more rural than urban, (2) their purpose is to preserve wilderness rather than domesticated nature, and (3) they are more concerned with landscapes or ecosystems than single species. For these reasons, they are much larger than botanical gardens. Nevertheless, in other respects they are similar: education and scientific

---

2 See https://www.bgci.org/garden_search.php (accessed 13 August 2014).
3 http://www.bgci.org/ourwork/international_agenda?sec=policy&id=international_agenda (accessed 20 August 2014). The Convention on Biological Diversity was opened for signature on 5 June 1992 at the United Nations Conference on Environment and Development (the Rio 'Earth Summit'); it is the most important document on biodiversity, and it has been signed by almost 200 countries.

research have central importance in both types of institution. Recreation and sport are permitted in nature parks, with some restrictions. Their aesthetics and atmosphere are greatly appreciated. The point of conjunction is probably the so-called 'English landscape garden', to be distinguished from the Italian or French ones because it is intended to exalt the spontaneous shapes of nature (Conan 1999), not squared hedges or framed flowerbeds, and the irregular and suggestive forms of shrubs and autochthonous trees are more respected. Thus, another dissimilarity arises: botanical gardens tend to nurture exotic plants, while nature parks aim to preserve local species.

This difference is important: Renaissance botanic gardeners were fascinated by foreign plants; modern nature park promoters are worried about the disappearance of entire local ecosystems, first in more 'civilized' countries, then in colonized ones. The history of nature parks is quite uniform; they arose in the second half of the nineteenth century in Europe and North America. They were conceived as antidotes to the environmental degradation resulting from urban and industrial expansion (and in the USA, large-scale hunting in wild areas as well). However, they assume slightly different meanings in old Europe and the New World. In the former countries, they tend to provide a field for scientific research, seen as a mission; in the latter, they are more monuments of national identity for people who lack old, magnificent buildings through which they can manifest their belonging (Piccioni 2013). Furthermore, they exhibit different scales: immense and inhabited only by indigenous minorities in America and Australia, smaller and incorporating agricultural activities in Europe. Nonetheless, in both cases a certain view of science and nature substantially as sources of normative values can be celebrated in the new temples that are the nature parks (Lackey 2003).

This inspiring idea is far from that of the botanical garden: the original local natural balance must be preserved against the drive to transform everything into assets typical of industrial or capitalist economies. This is the main cause of conflict with the usual stakeholders in wild environments. They are landowners who want their properties to be gated and available for the uses they prefer: the systematic harvesting of resources with modern practices. They may also be local authorities or ethnic minorities which claim exclusive jurisdiction over the land or seek to defend it against a central hostile administration. Finally, a plethora of nature users (such as anglers, hunters, hikers, off-road bikers and mushroom foragers) want freedom to organize their own activities within the protected areas.

There is disagreement over the type, extent and reversibility of changes to the natural balance of nature parks. Two philosophies have emerged since the beginning of their history: one sees nature as separate from human beings, and consequently demands that at least small portions of land be completely untouched by human activities; the other sees nature and human beings as so closely intertwined that protected areas are *laboratories* in which to experiment with a new equilibrium in the human–nature relationship (Brechin, Murray and Benjamin 2010). There is controversy between the two positions: the former (adopting the purist stance) maintains that supporters of the opposite view in the long run will justify interventions of every kind, even the more brutal ones; the latter (adopting a stance of human/nature compromise) accuse the purists of being ignorant of history because there is no area that has never been changed by human beings and 85 per cent of protected areas have a permanent human presence (Brechin, Murray and Benjamin 2010: 569). Furthermore, the purist conservationist position tends to lead to destructive conflicts with local residents and park users.

Some compromises have been reached to resolve these disputes. One is the strategy of the International Union for Conservation of Nature (IUCN, the main authority in the field) to create a *classification of parks* according to surface area, degree of protection from intrusion by human activity, and the presence of special or rare natural settings (endemic species). The IUCN has thus codified two types of reserves: – strict nature reserves and wilderness areas – and five types of protected areas, from national parks to protected areas with sustainable use of natural resources.[4] The other strategy has been to create zones with different degrees of protection within the same park (zoning as in urban planning). This principle is known as *gradualism* (see Section 4.6). With such polarized positions, the only way to advance is to graduate the levels of safeguards: the greater the complexity, the larger the number of protection levels.

Parks of different degrees of importance exist worldwide.[5] What varies is not only the degree of untouchability, but also the *institutional thickness*. The presence of a stable paid staff, inclusion within the national legal framework, and supremacy over other land use rules are all features that ensure a high institutional profile for nature parks. A crucial factor concerns the capacity of park authorities to engage in dialogue with people, be they visitors, local users or residents. Not only are these people involved in decision-making, they are also encouraged to be more active in the nature protection measures. Thus, their participation is political and practical at the same time. People *make decisions* about the nature park and *cooperate* to achieve its aims. This is a risky issue, especially in poorer countries, where such parks can be seen as a colonizing agency imposed on local people according to the desires and tastes of tourists from rich countries.

Over the last thirty years, procedural participation initiatives have been introduced into the institutional design of many parks (Héritier 2010). Typically, they take the form of consulting with stakeholders regarding park plans and programmes. On the other hand, authentic involvement involves three main aspects: (1) provision of financial means to support sustainable local economic projects, (2) formal inclusion of local interest groups and NGOs on the park board, and (3) the capacity to develop local knowledge and steer people's energies toward the aims of the park, for example through monitoring and surveillance. Nature storage

---

4 http://www.iucn.org/about/work/programmes/gpap_home/gpap_quality/gpap_pacategories/ (accessed 20 August 2014).
5 According to *ProtectedPlanet.net*, the online interface for the World Database on Protected Areas, the IUCN's Category II (national parks) comprises 467 in Europe, 688 in Asia, 335 in Africa, 2598 in America and 1,142 in Oceania. Of course, the surface area matters more than the number of parks. It is calculated that '12% of the Earth's land surface is contained within protected areas' (Brechin, Murray and Benjamin 2010: 563).

in situ therefore entails long and sensitive work with the local population, opening up several channels of interaction in an ideal triangle between the park authority, local interests and external visitors. The last two categories are often the missing links in active nature storage, which risks being conceived as the exclusive task of experts and functionaries.

The interactive or participatory method is also a way to discriminate the last form of nature storage we will consider: *seed banks*. Access to these by the public is greatly restricted. This aspect of seed banks is immediately apparent in the following description:

> A seed bank (also seedbank or seeds bank) stores seeds as a source for planting in case seed reserves elsewhere are destroyed. It is a type of gene bank. The seeds stored may be food crops, or those of rare species to protect biodiversity. The reasons for storing seeds may be varied. In the case of food crops, many useful plants that were developed over centuries are now no longer used for commercial agricultural production and are becoming rare. Storing seeds also guards against catastrophic events like natural disasters, outbreaks of disease, or war. Unlike seed libraries or seed swaps that encourage frequent reuse and sharing of seeds, seed banks are not typically open to the public.[6]

This quotation clarifies that seed banks are focused biodiversity storage devices, and that they are a response to two main objectives: to prevent the reduction of variability due to standardized agriculture, and to use variability as an instrument to increase the resilience of heavily perturbed systems. The quotation also confirms that a seed bank is an *abstract storage structure* in the sense stated in Section 5.1: an instrument for the precise selection of certain aspects of nature in their pure forms (germplasm) in order to preserve them against any environmental influences for an indefinite period.[7] This is an abstraction of any spatial or temporal contingency – a legacy of humanity that has absolute value, or at least an option value in the case of a catastrophe.

Other services, like seed libraries and seed swaps, can respond to more contingent needs. In this framework, we can imagine that seed banks are open only for exceptional events and to a restricted number of persons, presumably experts. In this sense, they closely resemble the IUCN's strict nature reserves, where access by the public is very limited. These banks eventually assume a sacred halo, according to Durkheim's idea of the sacred as inviolable. But there is another point: it regards the previously mentioned dialogical dimension of nature storage. If this

---

6 https://en.wikipedia.org/wiki/Seed_bank (accessed 20 August 2014).
7 A key challenge is maintaining the genetic purity and integrity of each stored seed species. This protection is not guaranteed in a botanical garden, and even less so in an open field, where plants of different subspecies may contaminate each other. Seed banks conduct periodic re-germination of seeds in a very pure, isolated environment to maintain their vigour.

kind of storage is so segregated, how can seeds be easily exchanged among individual structures and with potential users? On reading the *Strategy and Results Framework* of the network coordinator, the Consultative Group on International Agricultural Research (CGIAR 2010), one notes the special attention accorded to farmers in the global South; research on seeds should be beneficial to this category above all. Thus, the question of the relationship between seed banks and potential users becomes even more urgent, because very often banks are located in the North, while farmers in need are in the South.

There are about 1,500 seed banks throughout the world; the figure varies according to the source,[8] but the order of magnitude is likely to be correct. Like botanical gardens and national parks, the distribution of seed banks should be free of territorial bias. However, according to William Engdahl:

> The UN's FAO lists some 1400 seed banks around the world, the largest being held by the United States Government. Other large banks are held by China, Russia, Japan, India, South Korea, Germany and Canada in descending order of size. In addition, CGIAR operates a chain of seed banks in select centers around the world.
>
> (Engdahl 2007)

Engdahl's article is very critical of seed banks, which he sees as substantially the product of an alliance among the governments of Western countries, especially the USA, the largest plant breeders like Monsanto, DuPont, Syngenta and Dow Chemical, and international organizations such as the Rockefeller, Ford and Gates Foundations. The CGIAR itself was set up in 1972 by two of these foundations, Rockefeller and Ford. A slightly different picture emerges if we look at the fifteen seed bank projects selected by *Food Tank*, a blog of US AGProfessional in collaboration with the Science and Environmental Health Network (see Table 5.1).

It will be helpful to summarize the main features of these cases:

1 Table 5.1 lists projects rather than organizations, because a seed bank is often one among a range of activities usually, but not always, focused on agriculture.
2 The headquarters are spread throughout the world, although US locations predominate. Some organizations have more than one site, and some function as a hub for peripheral sites. North–South site partnerships are quite frequent.
3 The projects are very recent initiatives, while the supporting organizations may be twenty or thirty years old.

---

8 There were more than 1,000 in 2008, according to Debra Ronca, 'How seed banks work', http://science.howstuffworks.com/environmental/green-science/seed-bank4.htm (accessed 21 August 2014), and in 2012 there were 1,750, according to Longyearbyen, 'Banking against Doomsday', *The Economist*, 10 March 2012, http://www.economist.com/node/21549931 (accessed 2 February 2016).

Table 5.1 A selection of seed banks, their year of constitution and institutional profile

Name	Year founded	Headquarters location (institutional profile extracted from the organization's website)
AVRDC The World Vegetable Center	1971	Shanhua, Tainan City, Taiwan (non-profit institution)
Camino Verde, Living Seed Bank	2006	Concord, Massachusetts and Puerto Maldonado, Peru (US-based tax-exempt charitable organization)
Great Lakes Bioneers Chicago Seed Saving Initiative	2002	Chicago (local self-organized group of Collective Heritage Institutes/Bioneers)
Hawai'i Public Seed Initiative of the Kohala Center	2010	Kamuela, Hawai'i (independent, community-based centre, funded by Ceres Trust)
The International Center for Tropical Agriculture	1967	Cali, Colombia (non-profit research organization)
Louisiana Native Plant Initiative, Natural Resources Conservation Service (NRCS)	2011	Alexandria, Louisiana (initiative of NRCS, a federal agency of the US Department of Agriculture)
Man and the Biosphere Programme	1971	Paris, United Nations Educational, Scientific, and Cultural Organization (UNESCO)
Millennium Seed Bank Partnership	2000	Kew, London (initiative of Royal Botanic Gardens, an executive non-departmental public body)
Native Seed/SEARCH	1983	Tucson, Arizona (non-profit seed conservation organization)
Navdanya	1987	Uttarakhand, north India (registered as a trust in 1991)
New York City Native Plant Conservation Initiative	2008	New York City Department of Parks and Recreation, in partnership with Brooklyn Botanic Garden
New South Wales Seedbank, Australian Botanic Garden, Mount Annan	1986	Mount Annan, New South Wales (part of Royal Botanic Gardens and Domain Trust, a statutory body reporting to the Minister for Environment and Heritage)
Seed Savers Exchange	1975	Decorah, Iowa (non-profit organization)
Slow Food International	1986	Cuneo, Italy (non-profit international association)
Svalbard Global Seed Vault	2008	Longyearbyen, Norway (public-non-profit agreement; management entrusted to Nordic Genetic Resource Centre, financed by Nordic Council of Ministers)

Source: 'Update on the world's 15 largest seed banks', *Food Tank*, 1 August 2013, http://www.agprofessional.com/news/Update-on-the-worlds-15-largest-seed-banks-217990631.html (accessed 2 February 2016); author's own information added.

4   Most of them are initiatives by non-profit organizations. Some are public, depending on a national ministry, but with autonomy under a board of trustees.
5   Cases not mentioned include specialized seed banks like the International Rice Research Institute in Los Banos, the Philippines, and the International Potato Center in Lima, Peru.

The general picture is less negative than might be imagined: not only are these seed banks engaged with multinational seed breeders, some of them are explicitly oriented toward organic farming and openly opposed to industrial agriculture and genetically modified organisms. Furthermore, the geographical distribution of projects, and within them of different operative sites, is multilateral. In the end, the *terms of exchange* between seed banks and external stakeholders should serve as a good criterion for analysis. There are two main ways to regulate these exchanges:

- integrate in situ and ex situ biodiversity protection activities in the same organization;
- establish precise rules on access to seed banks by external actors.

The first method is clearly expressed by Native Seed/SEARCH (NS/S), a non-profit organization with a transborder operational field:

> NS/S utilizes a two-pronged approach to conserving crop genetic resources from the southwestern US and northwestern Mexico. *Ex situ* approaches involve conserving samples of crop seeds under frozen storage conditions, where they may remain viable (able to germinate) for long periods of time. We also utilize *in situ* approaches that support and encourage the ongoing relationship between people and plants through which both natural and human selection pressures continue to result in the development of new crop varieties – the same relationship between people and plants that produced the diversity present today.[9]

NS/S has integrated ex situ biodiversity (a seed bank) with direct management of a (conservation) farm, acquired in 1997 and used to grow and conserve native crops. Moreover, it promotes active dialogical conservation strategies focused on preserving the ecological integrity of the Sierra Madre Occidental in Mexico while meeting the cultural and economic needs of Tarahumara native communities by providing training and technical assistance and implementing model projects designed by local residents. Thanks to a cultural memory bank, NS/S seeks to 'combine the geneticist's concern for conserving unique traits of a crop with a folklorist's concern for conserving oral history about the crop'.[10] To complete

---

9   http://www.nativeseeds.org/our-approach/seed-bank (accessed 20 August 2014).
10 Ibid.

the vertical integration of biodiversity, NS/S has also created a sort of 'scattered arboretum', The Southwest Regis-Tree – a catalogue of remaining heirloom trees widespread in the region aimed at promoting their conservation and use. In sum, NS/S works across the entire in situ–ex situ continuum by creating a protected area in the Sierra Madre, instigating a botanical garden in the form of The Southwest Regis-Tree and the conservation farm, and, of course, setting up the seed bank. To this we can add the integration of the ecological dimension with the cultural one, since all the areas where NS/S operates are old lands of Native Americans.

The rules of the Australian National Seed Bank, a service under the auspices of the Australian National Botanic Gardens (ANBG), are an example of the second way to regulate exchange with stakeholders:

> The National Seed Bank can supply seed for approved research (not for profit) projects at other botanic gardens, universities and similar institutions by permit application. Seed is not supplied to private individuals. Applications for seed of species listed as threatened under the EPBC Act [Environment Protection and Biodiversity Conservation Act 1999] are assessed separately under an EPBC permit application.[11]

Private individuals are explicitly excluded. It is not clear whether private for-profit companies can receive the material. In the permit application, the ANBG specifies that 'any commercial use of the material, including essentially derived material, will require an additional legal agreement and negotiation of fees or royalties', so commercial use is allowed under certain conditions. One way to resolve this uncertainty is to restrict supply to non-profit organizations even if they make commercial use of the seeds. This is plausible because seeds are both universal goods (not appropriable) and have a functional value because they enable the cultivation of marketable crops. An NGO can legitimately organize the trade of autochthonous seeds in order to spread threatened local varieties, and at the same time earn revenue to pay its staff and fund the service.

Besides legal aspects,[12] the centrality of non-profit organizations is evident (Vernooy 2013: 5). It is true these may sometimes be Trojan horses for multinational breeder companies (Chapin 2004), but the situation is more complex than

---

11 http://www.anbg.gov.au/gardens/living/seedbank/index.html#seedbank (accessed 21 August 2014).
12 There are two legal issues: one concerns *property rights* over seeds; in this case, seed banks avoid the issue by claiming that they store materials whose owners are the depositors. The Norwegian Government website's frequently asked questions stated: 'Svalbard Global Seed Vault is not a gene bank, but a safety-storage for preservation of duplicate collections of seeds on behalf of genebanks'; http://www.popularmechanics.co.za/science/svalbard-global-seed-vault-seeding-future/ (accessed 5 February 2016). The other issue concerns the *limits on research*, which are unclear. Seed banks were created for scientific purposes; theoretically, they have no bias toward a particular type of research, even if the seeds are used for experiments on genetic modification. On more general issues, see Frison, Lopez and Esquinas-Alcazar (2012) and Halewood, López Noriega and Louafi (2013).

this. It depends greatly on how well they are able to combine their particular objectives with more general aims of the community. The *iron law of oligarchy* (Michels 1962) and *heterogenesis of ends* (Wundt 1892) are rules that apply to non-profit organizations as well (see Section 2.4). How they actually operate to promote universal and free biodiversity protection needs to be verified case by case. An interesting, even if spurious, case in this context concerns associations of professionals – farmers' organizations.

In Continental Europe, farmers' associations not only lobby, but also provide social and technical services. Their history has been politically fraught by their ambivalence toward authoritarian governments (Paxton 1997: 154ff.). Nevertheless, they have never limited their role to being pressure groups; rather, they provide self-organized services of various kinds, not least extension services (Jones and Garforth 1997). It is therefore disputable whether they should be excluded from access to seed banks. Some farmers' associations have been able to create their own community seed banks, demonstrating that they are stewards of the common good of biodiversity and should therefore be included in the governance of rural or wild areas.

Seed banks may be non-profit organizations, or when they are public bodies, may cooperate with non-profit organizations pursuing public objectives. Hence the non-profit associations are acknowledged as legitimate cooperators (see the references in Section 2.6). This occurs through registration on a public list after an accountability procedure. The organization is then allowed to manage the seed bank directly or to participate in one of its services. In conclusion, seed banks' operations are subject to two types of regulation: recognition of non-profit organizations as potential collaborators, and the application of strict access procedures for third-party utilization of seeds.

Environmental organizations are privileged candidates in balancing biodiversity protection and seed use. They usually have a democratic structure and conduct commercial activities without making any profit. Moreover, they are considered more efficient in the protection of endangered species (Shandra et al. 2009). If they respect these conditions, they are not only recognized as rightful members of biodiversity protection networks, but can also directly manage reserves, botanical gardens and seed banks.

In spite of their ideal position, the influence of non-profit organizations in seed bank management is less marked than it is in the promotion of other forms of biodiversity storage. In the case of protected natural areas, such organizations are sometimes on the board, but they are rarely the principal manager (see Table 5.2). This applies to reserves that are much smaller than national parks, and generally created to protect a single species or habitat. Botanical gardens, as we have seen, are more often managed by a public body with substantial autonomy. It is true that animal sanctuaries also have significant involvement of non-profit, if not private, organizations, but this seems to rely on the scale of land use. Large protected areas require publicly owned land or land under special jurisdiction; private companies or environmental organizations do not have the financial and legal means to administer extensive tracts of land.

150 Long-term life storage

Table 5.2 Main features of three forms of biodiversity storage

Storage form	World network	Convention or treaty	Institutional profile
Botanical gardens	Botanic Gardens Conservation International	Convention on Biological Diversity (opened for signature on 5 June 1992)	Scientific public bodies
Nature parks	International Union for Conservation of Nature	Ramsar Convention on Wetlands, 1971; Convention Concerning the Protection of the World Cultural and Natural Heritage, 1972	Dedicated public bodies, NGOs for reserves
Seed banks	Consultative Group on International Agricultural Research	International Treaty on Plant Genetic Resources for Food and Agriculture, 2004	Foundations, trusts, non-profit organizations

Despite differences in their ages, the institutional similarities among the three types of biodiversity storage are striking. In all cases, there are an international network and one or more international treaties. The world network does not have authorization and sanctioning powers: it mainly provides coordination and information services; it rarely establishes criteria and standards, which would give them some cognitive power to frame issues. The network supports official bodies like the United Nations in the drafting of international treaties and the implementation of rules. The international organizations carry out numerous projects, from which they obtain most of their funding, donations being insufficient. This *dependence on projects* is becoming a problem for organizations of this type, which are being driven to focus more on seeking sources of finance than attending to their original missions.

A parallel can be drawn between the worldwide organization of biodiversity and human genome globalization. According to Pellizzoni:

> The human genome [...] is global in a threefold sense: technological (online database accessible all over the world); scientific (through access and distribution of data, scientific knowledge can be shared worldwide) and economic (data in the genome database are connected to patent databases, with consequent restriction of information-sharing).
>
> (Pellizzoni 2015: 28)

The first two features – technological and scientific – are common to biodiversity storage forms: accessible and systematic biodiversity databases are maintained by an information and communication technology-based scientific international network. The difference consists in both the codification of knowledge and the possibility to patent biodiversity, at least in the form of ecosystems. Ecosystems, or in situ biodiversity, are difficult to frame in discrete units of analysis. The reduction of the ecosystem to an object is hampered by various cognitive problems, such as how many species there are, how they interact and what the borders are. For these reasons, it is very difficult to measure the *absolute* degree of biodiversity in a protected area. It has been shown that the presence of an in situ storage structure assures the safeguarding of a *relative* degree of biodiversity (Brechin, Murray and Benjamin 2010). Similarly, the botanical garden – the intermediate category between in situ and ex situ – certainly has the capacity to protect biodiversity, albeit restricted to single plant varieties.

More controversial is the role of seed banks, which do great deal of work with genetic agricultural material and are at risk of economic exploitation. We have seen that such structures defend universal access to seeds both by entrusting their custody to non-profit organizations and by enforcing a strict procedural process consisting of highly formalized applications, evaluations and concessions. In both cases, there is an attempt to ensure the widest right of entry through flexibility, discretion and case-by-case assessment. Nevertheless, the question of whether equal access to biodiversity can be given to poorer farmers while ensuring adequate protection of seed specificity is still an urgent one. According to the Indian Centre for Development, the answer is largely positive in those cases where a *community seed bank* has been created:

> Seeds are given free of cost to members of a seed bank. Any one from the community can become a member by paying a nominal annual fee. The member then sows the seed, harvests the crop, and later returns to the seed bank twice the quantity he received to replenish the store. The seed bank also works on seed treatment, seed selection, maintaining a record of needs, and planning for the next season. The seed banks are managed by women's groups.[13]

It should be noted that the bank does not solely catalogue and conserve exemplars of seeds; it also stores and lends massive quantities, like a financial bank. Access is reserved only to members, but the fees are very low, and furthermore, marginalized sectors of society, such as women, have a central role. Community 'seed banks and seed swaps exist all over the world and are continuing the precious heritage of seed'.[14] Even in countries with highly industrialized agriculture, small-scale schemes exist (Rights of Mother Earth 2014) under a variety of names and frames – 'In the "West," many so-called seed savers groups, associations and networks are made up mostly of (hobby) farmers and gardeners' (Vernooy 2013: 6) – and without a strict 'place-bound social group' as in the South.

Problems arise when this agricultural system begins to export goods beyond local borders, in which case a need for cataloguing of varieties arises, the justification being defence of the intellectual property rights of those who have spent

---

13 http://base.d-p-h.info/en/fiches/dph/fiche-dph-8060.html (accessed 28 August 2014).
14 http://www.seedsoffreedom.info/about-the-film/frequently-asked-questions/ (accessed 29 August 2014).

such a long time selecting and preserving the genetic information contained in the local seeds: 'Several of these free trade agreements contain clauses relating to intellectual property rights which sometimes ask countries to reinforce their legislation on the intellectual protection of plant varieties' (De Schutter 2011).

International trade rules also markedly favour the protection of private property in the case of immaterial goods like knowledge. According to this philosophy, open-source seed supply does not reward the time and expertise a single farm or company may have expended, and hampers investment in further research. However, restrictive use of intellectual property rights rules reduces access to seed resources for people and countries with low *cataloguing capacity* (Kloppenburg 2010). The introduction of controlled seed storehouses appears to be an acceptable compromise to overcome this dilemma. A controlled seed storehouse can organize seed trading in a flexible manner, can serve as a learning centre for seed selection and classification, and can filter potential users through compulsory membership and strict procedures. It can become, as has happened in various historical periods, a *social bank* whose role is to accumulate wealth in the hands of many (Weber 2012).

In conclusion, seed banks, as well as nature parks and botanical gardens, are concentrations of biodiversity and human knowledge that: (1) counterbalance the extraordinarily concentrated market power of the major seed companies, (2) reduce *bioprospecting* and *biopiracy* by supervising seed exchanges, and (3) maintain a compromise between the right to privately appropriate genetic resources and the right to exchange commons freely. Such biodiversity storage structures not only conserve biodiversity, they actively exchange exemplars of it. This occurs within ambiguous legal frameworks and against consolidated interest groups. But the challenge seems clear: enlarging the biodiversity storage forms beyond the official research system and the small niches of 'alternative' peasants.

5.2 Beyond land sparing and land sharing

The previous section described different ways to store biodiversity according to the degree of isolation from the environment, from in situ protected areas to ex situ seed vaults. Storage in those cases was a means to conserve biological diversity. The criterion was to maintain exclusive areas for conservation of non-human life forms. In the Anthropocene epoch, the large majority of areas are dedicated to human ends. Habitation and production are the main land uses. This is an obvious statement, yet it explains very well why exclusive conservation zones are so small and few in number, and why the largest of them, such as national parks, must reach compromises with many other land uses. This is precisely the point: because most of the world's land is subject to compromises among many uses, the storage of precious biodiversity has to be conducted within these mixed areas.

The example of the Trentino bear in Box 5.1 is eloquent: in Europe, even in mountainous areas, it is impossible to find exclusive zones that are sufficiently extensive to satisfy the bear's needs for nutrition and reproduction. Some zones must necessarily be mixed – that is, areas co-inhabited by human beings and bears. As this account shows, this leads to numerous problems, but there are no alternatives in densely populated zones unless people are forcibly excluded. In other words, nature storage systems must be tailored to a variety of environments, even to the extent of treating certain towns as biodiversity sanctuaries.

Box 5.1 The Trentino bear destined for an animal sanctuary?

Trentino is a small province in the Italian Alps, once part of South Tyrol. In 1999, its local authority decided to introduce some bears from Slovenia. The term 'introduce' is not strictly correct, because the process was initially termed *rinsaguamento* – 'blood or genetic renewal'. According to some local sources, at the time of introduction there was a residual population of autochthonous bears in the some valleys in western Trentino. The aims of the project therefore included giving new vigour to the few old bears still surviving. It should also be borne in mind that the local population were hostile to bears because folk memories of them as the enemy of humans.

Nevertheless, the project was launched in the province's Adamello-Brenta Park, and it was successful. The introduction came about in several phases. The Slovenian bears adapted very well to the new environment, rising to a population of about fifty. Of course, they did not remain within the park's boundaries, and moved hundreds of kilometres away from the original area. Some arrived in Germany, Switzerland and the nearby Italian regions of Veneto and Lombardia. Until 2014, the problems of this spread were limited to the killing of some livestock, the need for rapid reimbursement for bear-caused damage, and baseless fears among tourists.

In the summer of 2014, a mushroom forager was attacked by a female bear with two cubs. Fortunately, the clash was not fatal, but the repercussions were serious. For the first time since the introduction, there had been a violent encounter between human and animal. Infuriated groups with opposing views held rallies. Animal rights activists proclaimed the right of bears to defend their offspring, while local people declared that their right to safety was threatened. The response by the provincial authorities to date has been to extend monitoring as far as possible through the use of radio collars. The idea is that knowledge of bears' habits will help to prevent damage and possibly modify their feeding areas. Troublesome bears can be confined within fenced areas, or as the extreme solution, killed.

Since the attack, opposition to the in-situ presence of bears has increased. The local authorities fear excessive reactions by local people and harmful consequences for tourism. The most likely solution will be a compromise between in situ and ex situ: an animal sanctuary. An extensive fenced area for bears already exists within Adamello-Brenta Park, and it could be enlarged. This is a classic compromise: bears are present, but not in ugly cages.

Overall, sanctuaries have the advantage that they can be visited without danger; this is what people like the most: admiring living nature and charismatic animals, but in controlled situations. This also explains the reasonable success of animal refuges or sanctuaries in Western countries despite heartfelt pleas for wilderness.

154 *Long-term life storage*

The dilemma is well conceptualized in the debate on land sparing versus land sharing. The following quotation states the matter clearly, connecting it to our general issue of storage as a way to respond to a new era of scarcity:

> As the demands on agricultural lands to produce food, fuel, and fiber continue to expand, effective strategies are urgently needed to balance biodiversity conservation and agricultural production. 'Land sparing' and 'wildlife-friendly farming' have been proposed as seemingly opposing strategies to achieve this balance. In land sparing, homogeneous areas of farmland are managed to maximize yields, while separate reserves target biodiversity conservation. Wildlife-friendly farming, in contrast, integrates conservation and production within more heterogeneous landscapes. Different scientific traditions underpin the two approaches. Land sparing is associated with an island model of modified landscapes, where islands of nature are seen as separate from human activities. This simple dichotomy makes land sparing easily compatible with optimization methods that attempt to allocate land uses in the most efficient way. In contrast, wildlife-friendly farming emphasizes heterogeneity, resilience, and ecological interactions between farmed and unfarmed areas. Both social and biophysical factors influence which approach is feasible or appropriate in a given landscape.
>
> (Fischer et al. 2008: 380)

Land sharing and *wildlife-friendly farming* are synonymous (Tscharntke 2012). The important points for our argument here are, first, that there is no best model in absolute terms. Intersections with human activities and among wild species are different; consequently, the conservation of richer ecosystems can be attained both through land sharing and land sparing, according to local circumstances (Hodgson et al. 2010). In highly urbanized areas, like the Italian Po Valley or the Ruhr basin in Germany, there is no alternative to land sharing. Second, our recurrent dichotomy between storing and networking is further refined: the preference for one or the other also depends on the urban structure, as Figure 5.2 illustrates. The presence of a *primate city*, disproportionately larger than any others in the urban hierarchy of the country (Jefferson 1939), probably induces a sharp separation between over-populated and wild/rural areas, leading to a land sparing pattern.

Furthermore, it is interesting to consider two other types of interconnections between farming and biodiversity: Mediterranean and mountainous areas. The former are more similar to the Western Australian example in Figure 5.2: high urban concentrations on the coasts, and wilder internal areas, often with semi-arid climates. In these cases, the opportunities for land sharing will depend on whether the irrigation systems utilize farm pond networks (see Section 3.4), which support the development of wild species. When mountainous areas are not affected by highly industrialized agriculture (such as fruit growing in South Tyrol), they exhibit a pattern closer to land sparing: a valley floor with the maximum concentration of buildings and infrastructures, and mountain slopes covered with woods because of the abandonment of farmland. The repercussions of the spread of woodland in some mountain areas

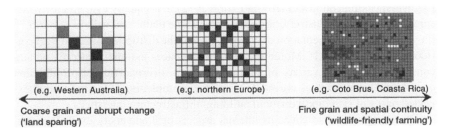

Figure 5.2 Conceptual model of the continuum of scales at which biodiversity conservation and agriculture can be integrated

Source: Fischer et al. (2008).

are dramatic: landscape change, increasingly wet weather and the proximity of wild animals. This provides an impetus toward the land sparing pattern.

Third, the two biodiversity conservation models recall strong cultural assumptions about human preferences: 'Different scientific traditions underlie the endpoints of the continuum [see Figure 5.2]. These traditions influence how the task of balancing biodiversity and agriculture is conceptualized and accomplished, but they are rarely drawn out explicitly' (Fischer et al. 2008: 381). For example, in forestry there are three different tree cutting patterns: 'single tree selection, group selection, and strip clear cut' (Rocky Mountain Forest and Range Experiment Stations 1963: 2). Each of them presumably has its own capacity to maintain biodiversity over time and to tackle other problems, like CO_2 absorption (Martin, Newton and Bullock 2013). Nonetheless, landscape preferences are important as well; sharply squared cleared plots in the middle of a forest are not visually appreciated, nor are long strips cleared of trees to create room for power lines or ski trails. Even if such open spaces increase biodiversity, they are perceived as wounds on the body of the mountain. At the same time, a very uniform wooded landscape is less preferable than a varied one consisting of pastures, hamlets and scrubland – a typical Alpine setting (Schirpke et al. 2013).

Moreover, biodiversity storage assumes many forms and responds to a mix of criteria: the increasingly invoked functionality to human life (ecosystem services), sublime aesthetic tastes, the availability of resources for future generations, and finally, the landscape's capacity to furnish an image of *human relationships with the environment*. Undoubtedly, the preference for land sharing or sparing stems from a need for a neat representation of how people want to mix with nature. Landscape becomes a metaphor of people's preference for *promiscuous* or *agonist* relationships with wild environments. Biodiversity conservation, like many other storage practices, helps to distinguish symbolically and socially. We have seen this in food storage (Section 2.5). It also emerges in the management of water (Cohen and Harris 2014) and energy (Osti 2012a). It is now evident for more natural landscapes.

Having accepted the cognitive function of biodiversity storage forms in terms of relations with agriculture, we must acknowledge there is also a spared/shared combination in residential and industrial activities. For example, it is clear how

156 *Long-term life storage*

much scattered developments of residential properties in forests are appreciated in English-speaking countries (Troy and Voigt 2012); it is equally evident how widespread small-scale industrialization has affected the plains and valleys of northern Italy, provoking fragmentation of ecosystems and the disruption of ecological corridors (Marchetti 2002). Moreover, peri-urban spaces assume a specific pattern consisting of a great variety of exotic species, a disproportionate proliferation of invasive birds (such as crows) or trees (for example, black locust), and unexpected oases of rich biodiversity in building–infrastructure interstices.

The complexity also concerns internal aspects of biodiversity itself. It is subject to many independent classifications which produce a 'honeycomb' structure, rather than a simple continuum between low and high biodiversity (see Figure 5.3). Socio-political dimensions are important in the definition, use and protection of biodiversity (Escobar 1998; Brechin et al. 2002; Brown 1998). Claims for *subsidiarity* in protected areas provide the best examples of the quest for new centre–periphery relations (Borrini-Feyerabend, Kothari and Oviedo 2004: 91).

This socio-environmental complexity illustrates how biodiversity storage is in serious trouble. The alternative between separation (sparing land) and mixing (sharing land) is too inflexible. Indeed, nature protection agencies have devised at least two more fine-tuned strategies. One is the creation of *buffer zones*; the second is the creation of *green islands or belts* within the most urban-industrial areas. The former respond to the principle of *graduation* and of compromise; there is an acceptable transition from most-used to forbidden zones with, in the middle, those allowing recreational uses like hiking, hunting or bird-watching. The latter

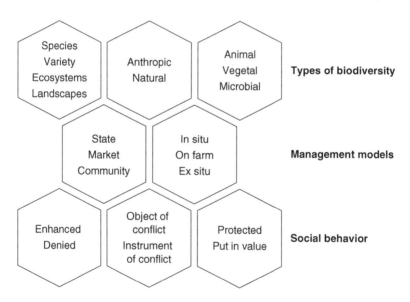

Figure 5.3 The honeycomb of biodiversity: socio-cognitive dimensions of biological diversity

Source: Carrosio (2015a).

are only superficially similar to the land sharing model: they are not simply a mix of agricultural and greener landscapes. Theoretically, this approach is called *re-entry*, and consists of reproduction of the system/environment distinction within the system itself (Wolfe 2010: 238).

In a situation where biodiversity is a source of difficulty because of its complexity and uncertain evolution, it is preferable to include it within likewise complex urban or agro-industrial contexts. Urban parks, vegetable gardens, green belts, and even multi-storey agriculture, university campuses or factory ponds, are nothing but semi-natural environments under the strict control of a human residential system. They resemble biodiversity islands, not only circumscribed, but also penetrated by highly artificial structures. Consider, for instance, the artificial sealing of urban pond bottoms. These are similar to silos used to store grain or water. Symbolically, they are open-air laboratories where experts and non-experts can explore nature. They are modern botanical gardens whose task is to celebrate the adaptation of nature to human ideals.

The re-entry mechanism can be applied to nature conservation agencies as well. Parks, forest corps and nature sciences institutions conduct their activities fully immersed in semi-wild areas. But within them they create structures that help to simplify their ecosystems. In fact, they build eco-museums, observatories, guided paths and so on which allow the fruition and understanding of nature. In other words, they create further system/environment distinctions within protected areas. The physical forms of these systems can be very variable – from small buildings to simple trails. Nevertheless, the purpose is the same: to provide improved opportunities for appreciating stored – and, hopefully, protected – biodiversity. Giving this wishful thinking material form cannot be a casual process: the observation and exploitation of ecosystems lead to their consumption, so it is crucial that the nature protection system observes itself by means of self-monitoring mechanisms, re-entry among them.

5.3 The mutual exchanges between biodiversity and storage

Thus far, we have dealt with various aspects of biodiversity storage: how organizations provide special places for storing biodiversity, and how the primary activities of agriculture, forestry and livestock raising can be coupled with the conservation of semi-natural areas. In addition, we have explored how industrial and residential activities are interconnected with biodiversity storage attempts. The third and final point concerns the mutual exchange between biodiversity and the storage of other substances that are more or less useful to human beings. It can be expressed in two ways:

a   biodiversity storage can reduce some environmental problems resulting from human activities – it can provide ways to store potentially harmful substances such as nitrates or CO_2 in trees, soil and water;
b   typical human storage activities can actually increase biodiversity – for instance, artificial water basins, which can provide a habitat for a wide variety of species and ecosystems.

158 *Long-term life storage*

There is a large body of literature on the capacity of different ecosystems to absorb CO_2. Forests are generally considered the primary sites in this respect, but experts are far from reaching consensus about their utility: 'The best way to manage forests to store carbon and to mitigate climate change is hotly debated' (Bellassen and Luyssaert 2014: 153). The same authors declare that the topic is a sort of 'gamble'. More prudently, Hicks et al. (2014: v) state: 'there is *established but incomplete evidence* supporting the link between species richness (and diversity) and forest carbon sequestration' (italics in the original). High uncertainty depends on the following factors:

a biodiversity multidimensionality – there are three main criteria for assessing this: richness of species, number of threatened species and number of restricted-range species;
b the scale of processes – the minimal area required for conservation of a species;
c unknown threshold effects – popularly known as 'the butterfly effect';
d difficulties in calculating trends compared to the initial biodiversity stock.

Even considering all these levels – individual, species and ecosystem – and further parameters (such as intactness), biodiversity has such a variable composition that any results will be unreliable. Nonetheless, storage practices are invoked in three main domains:

1 *carbon capture and storage* (CCS) – capturing fossil fuel CO_2 emissions and depositing them in large (usually geological) structures;
2 creation of *ecosystem carbon storage pools*;[15]
3 *synthetic biology* – providing CCS and other ecosystem services.

Biodiversity can play a role in the second and third types of carbon storage. The main hopes rest on ecosystem carbon storage pools, especially forests, because of the many co-benefits that derive from growing trees.[16] The storage practices perspective can help in evaluating these functions; for example, a crucial issue is *duration*: how long-term is the carbon sequestration of forests? Radical clear-felling of trees for energy purposes, such as is practised in short-rotation forestry, makes the sequestration of CO_2 very limited. Furthermore, the natural decomposition of

---

15 They are classified as urban and quasi-urban, farming and forest ecosystems, which in turn can be subdivided into 'five carbon storage pools: living trees, down dead woods, understory vegetation, forest floor, and soil' (Onaindia et al. 2013: 4), in addition to which there are coastal and open sea ecosystems.
16 It is impressive to list forest ecosystem services. In addition to CCS, Hicks et al. (2014) distinguish 'other regulation services', such as productivity, soil erosion control, soil fertility and nutrients, pollination, seed dispersal, fishery enhancement, disease prevention, biological control, water regulation and quality control, protection from natural hazards, climate regulation, pollution control and 'provisioning services', such as timber and non-timber forest products – that is, food, fodder and medicines.

trees and other vegetation, which is highly appreciated by conservationists, liberates previously stored CO_2. Replanted forests seem to function better for carbon sequestration than for biodiversity conservation, as it takes at least hundred years to return to climax (Gilroy et al. 2014). In any case, to buy time, long-term atmospheric carbon storage in trees is a good strategy: there is a mitigation effect on greenhouse effects, and in the longer term, it is possible that other solutions may arise (Bellassen and Luyssaert 2014: 155). In one way or another, time works in our favour; but the issue of selecting the best-performing species remains. What mix of species and habitat provides the longest-term carbon sequestration?

One solution involves the conservation of mangroves, seagrasses and salt marsh grasses, since this coastal vegetation – dubbed 'blue carbon' – sequesters carbon up to a hundred times faster and more permanently than terrestrial forests (Pellizzoni 2015: 1). The ranking is definitively in favour of forests rich in water: the Amazon rainforest absorbs a quarter of the 2.4 billion metric tons of carbon captured each year by the world's forests (Kintisch 2015). However, counter-intuitive processes operate together with climate change. The darker forests absorb more CO_2 than tundra or savanna, but retain more heat, thus contributing to global warming.

It is consequently very difficult to establish a unique correlation between biodiversity and carbon storage. The most mature analyses establish territorial differences:

> Deforestation is a main driver of climate change and biodiversity loss. An incentive mechanism to reduce emissions from deforestation and forest degradation (REDD) is being negotiated under the United Nations Framework Convention on Climate Change. Here we use the best available global data sets on terrestrial biodiversity and carbon storage to map and investigate potential synergies between carbon and biodiversity-oriented conservation. A strong association (Spearman's rank = 0.82) between carbon stocks and species richness suggests that such synergies would be high, but unevenly distributed.
>
> (Strassburg et al. 2010: 98)

The unbalanced distribution of ecosystems able to provide both carbon and biodiversity storage has policy implications:

> Many areas of high value for biodiversity could be protected by carbon-based conservation [REDD-type measures], while others could benefit from complementary funding arising from their carbon content. Some high-biodiversity regions, however, would not benefit from carbon-focused conservation, and could become under increased pressure if [only] REDD is implemented. In particular, if a large-scale carbon-focused REDD mechanism is implemented, biodiversity-rich and relatively carbon-poor regions could suffer from a double conservation jeopardy, with conservation investment diverted away from them, and human pressure redirected toward them, as carbon rich areas become the focus of conservation efforts. Areas potentially at risk include

some that are widely recognized as global biodiversity conservation priorities such as the Brazilian Cerrado, the Cape Floristic province, and the Succulent Karoo.

(Strassburg et al. 2010: 101)

Even more uncertain and contested is the contribution of synthetic biology to biodiversity and carbon storage:

> Synthetic biology is a form of extreme genetic engineering that adds manufactured genetic parts (such as synthetic DNA, synthetic ribosomes or synthetic RNA) to a living cell in order to *hijack* the workings of the cell for industrial uses. Adopting engineering principles, researchers attempt to create modular *genetic parts* or *biobricks* that can be easily snapped together to create more complex genetic *programmes*.
>
> (ETC Group 2010: 2)

Synthetic biology is a genetic programme which stores information that will not be contaminated by other living organisms. According to Schmidt (2010), it should be considered an ultimate warranty of biosafety in a world continuously threatened by natural and artificial pathogens. This feature should be useful for those storage practices that rely on the conservation of a living process – fermentation, for example– and must isolate possible causes of degradation – to continue the example, the agents that transform wine into vinegar.

Synthetic biology concerns not only biodiversity conservation but also its re-establishment: 'Extinction might not be forever if synthetic biologists and others pursue their proposals to use advanced genetic engineering techniques to save endangered species and return extinct ones' (Redford, Adams and Mace 2013: 1). This perspective is fiercely contested in two fields: 'the potential invasiveness of these artificial life forms in the wild', and the massive deterioration of natural resources once these organisms are used at industrial scale in poor ecosystems or for agriculture wastes (ETC Group 2010: 3). The promise is to increase productivity in the food and energy sectors and to find biological cleaners of the main pollutants. Both processes will be 'biological' – that is, performed only by living organisms – and are therefore considered better for the same reason that organic food is preferable to food treated with chemicals.

The problem is first philosophical, and then technical: it is difficult to assess the many positive and negative aspects of synthetic biology (International Civil Society Working Group on Synthetic Biology 2011). The evaluations are:

> hard to frame because it is difficult to identify the right counterfactuals or alternative futures to compare with those underpinned by the new technology. It seems inevitable that synthetic biology will be a major factor in affecting the future. That future world will not be a slightly older version of the world that we currently inhabit.
>
> (Redford, Adams and Mace 2013: 3)

Storage as a cognitive frame can help to tackle these controversies. Storage is not part of the intrinsic design of nature, which continuously destroys and reproduces itself in a variety of ecosystems and cycles. Storage, even of biodiversity, has a special cultural purpose that emerges clearly if we consider the desire to return extinct species to life. This has parallels in the aspiration to maintain, or even reproduce, past forms of social life, as in open-air living museums. Conservation or restoration of the past is a typical human activity which pertains to the field of socially constructed preferences. The issue thus becomes a matter of science policy; the question is whether to invest so many resources in synthetic biology – which is more coherent with laboratory procedures of verification, but yields very remote results – or whether to invest more in in situ ecological research, which acquires less rapid scientific evidence, but has less obscure consequences and side-effects.

In situ research into biodiversity's capacity to provide ecosystem services is a case in point that is coherent with our storage perspective. It is a consolidated result that stagnant water ecosystems are able to store nitrogen from human activities:

> Over the last 20 years, 59 experiments have quantified how the richness of plants and algae influence concentrations of inorganic nitrogen in soil or water. Of these, 86% have shown that the concentration of nitrogen decreases as biodiversity increases – by an average of 48%.
>
> (Cardinale 2011)

We know that nitrogen is a problem due to bad wastewater treatment and heavily fertilized agriculture. Both release, on the ground and beneath it, quantities of nutrients that above some thresholds have negative effects on human health and on the environment itself (Grizzetti 2011). The damage is also economic if we consider, for example, the case of the algae overgrowth (eutrophication) in the Adriatic Sea, which leads to problems in the use of beaches for tourism.

It is noteworthy that carbon and nitrogen storage are studied together in search of win-win solutions:

> Collectively, our findings indicate that changes in plant species and functional group richness influence the storage and loss of both C and N in model grassland communities but that these responses are related to the presence and biomass of certain plant species, notably N fixers and forbs. Our results therefore suggest that the co-occurrence of species from specific functional groups is crucial for the maintenance of *multifunctionality* with respect to C and N storage in grasslands.
>
> (De Deyn et al. 2009: 864; emphasis added; see also Cong et al. 2014)

However, the capacity of ecosystems with high biodiversity to absorb both atmospheric carbon and nitrogen is variable and subject to side-effects or trade-offs: 'Peatlands contain twice as much carbon as all forests combined, while only covering 3% of the Earth's land space' (BiodivERsA 2014: 1). But if invested with enhanced nitrogen deposition, they release that carbon very rapidly, thus losing their valuable function as carbon sinks. For this reason, research is seeking to discover the mechanisms and substance dosages which favour the cumulative beneficial effects of ecosystems (Rillig et al. 2007). This research combines in situ approaches with those used in laboratories in order to match ecological analysis as much as possible with biochemistry, which studies essential combinations of animate and inanimate elements. Biodiversity adds to the storage perspective, based on the coordination of time and space (see Section 1.2), a further criterion: the appropriate combination of elements. If carbon or nitrogen storage demand extremely large areas and long periods to be effective, the same processes can be improved through a calibrated combination of artificial and natural factors.

This appears to be further confirmation that every form of storage is a *cultural appropriation of natural processes*. It emerges clearly when we look at how storage practices modify ecosystem biodiversity. Traditionally, the storage of food and water provoked the proliferation of an unpleasant biodiversity consisting of alien or autochthonous parasites and weeds – a wide range of harmful species both for people and crops. The two paradigmatic examples are rats in granaries and mosquitoes in water reserves. Chapters 2 and 4 illustrated how storage practices have involved constant measures to counter such natural disease vectors.

The perspective can be reversed by looking for cases where the need to store things useful for humans needs has had a positive effect on biodiversity. In broad terms, agriculture and pasturage can be seen as ways to differentiate the environment. A frequent case cited in the context of the Alps, but probably valid for all mountainous areas, is the abandonment of high-altitude pastures. These were habitats variably colonized with sheep and cows across the centuries. After the Second World War – this is mainly the case in Italy – many pastures were abandoned, and woods rapidly re-colonized the resulting open spaces. According to many experts (Spehn, Liberman and Korner 2006), the combination of forest, scrub and pastureland represented an extraordinarily biodiverse ecosystem for both vegetation and animals. This argument is frequently invoked in terms of the survival of bears and capercaillies (*Tetrao urogallus*) as well. These arguments go hand in hand with calls for a return to old agricultural and pastoral practices. The irony is that such a return, when it happens, is frequently practised by outsiders – migrants from other countries. Hence, biological diversity is enriched by cultural diversity, thus reaffirming that it is impossible to restore the past, and that new mixes of nature and culture will continue to advance.

Another macro case concerns hydro power basins. Artificial pools certainly lead to ecological changes in the depressions where they are created. The resultant increases in fish species and aquatic vegetation can be controversial, especially in the global South (Ziv et al. 2012), while in the global North, the conflict over ecological issues is less marked (Pringle, Freeman and Freeman 2000). The difference seems linked to the *more frequent public monitoring* of basin in temperate area basins, which dilutes the opposition through the use of multicriteria analysis and stimulates adoption of concrete remedies like installing fish ladders alongside dams.

When the production of hydro power is at its most sophisticated (in the case of pumped water plants: see Section 4.1), the water basins must be kept very clean. Waterborne debris and heaps of it where a variety of species can find a habitat cannot be tolerated because they impede water flow or reduce the basin's capacity. It is difficult to imagine a flourishing of biodiversity in such situations. Nevertheless, some artificial basins, especially very large ones, certainly provide venues for biological and cultural diversity, as well as open-air activities like fishing.

Evaluation of how dams intertwine with biodiversity must be conducted at basin level as well. Even though the dendritic structure of the water stream imposes a clear direction on the flow, a number of species are able to swim against the current. These are called diadromous fish – species that spend part of their lives in fresh and part in salt water. The degree and patterns of mobility are important for calculating their chances of survival in the presence of artificial barriers. A group of US researchers has designed a set of riverscape principles to guide the ecologically sustainable development of river basins for hydro power:

> (i) within a large river basin, concentrate dams within a subset of tributary watersheds and avoid placing hydropower facilities on a downstream mainstem, (ii) disperse freshwater reserves among remaining tributary watersheds, (iii) ensure that habitat between dams will support and retain biological production, and (iv) formulate spatial decision problems at the scale of large river basins.
>
> (Jager et al. 2015: 815)

These principles warrant some comments in order to draw out useful information not present in the authors' synthesis. First, a large-scale approach, in this case at watershed level, is necessary, both for analysis and for decision-making. Second, the ranking of aims through techniques like the Analytic Hierarchy Process (Lombardi 2015) remains the best approach, but with the complication that not only must biodiversity and energy production be considered, but also other water objectives, like drought prevention and maintaining freshwater reserves. Third, the relationship between storage and network is once again variable – it cannot be encapsulated in one model. The rule of (bio)diversity also applies to hydro power plants: it is preferable to avoid installing the same types of plants on all the tributaries, to serve different objectives. The relative specialization of single tributaries increases biodiversity, including that generated by human activities, and it reduces risks in the case of extreme events. Fourth, the typical American attitude toward change allows decision-makers to contemplate the demolition of some dams or their transformation into impoundments (damless hydro power facilities) – this room for manoeuvre gives flexibility to bottom-up participation and agreements. This contrasts with Europe, where attitudes are much more conservative toward nature and artificial structures.

Finally, biodiversity induced by recreational activities must be considered. Fishing and hunting are probably the most contentious issues in the mutual exchange between storage and biodiversity. *Fish ponds* and *hunting reserves* are cases in point. This controversy can be formalized according to the source/sink dualism (Hansen 2011) – a frame related to buffer zones as well (see Section 5,2). Some delimited areas or basins can be protected to promote the reproduction of game species. Such areas or ponds must have a good-quality environment (source) in order to provide vigorous animals able to migrate to the places where they will be hunted, and where environmental standards are usually lower (sink).

The ambiguity of this position has been evident since the times when lords created exclusive hunting grounds. Hunters and anglers promote the setting up of reserves in order to increase the game population. The biodiversity of the source areas is promoted, in that it creates the best conditions for the rapid and abundant proliferation of game species. This selection of eco-objectives can be dangerous for species that are not complementary to game animals. A wood with a reserve for hunting deer surrounding it may be destroyed because of deer over-population. A typical problem with many protected areas is the overwhelming presence of wild boar, which no longer have natural enemies. The introduction of antagonists like wolves raises many other problems, such as those mentioned earlier for bears (see Box 5.1). Thus, there is not only a need to deal with unpleasant or unwanted biodiversity, but also an imperative for constant management of the combination of species. Science can provide answers through simulations of the spatial diffusion of species according to a variety of ranked criteria. However, this is difficult because of many inter-species trade-offs and intrusion by alien species, and also because the human preference for biodiversity combination is unclear or unspoken.

5.4 Mosaics of biodiversity

The first issue to address in these conclusions is whether the storage perspective has contributed something new to knowledge about society. Storage is synonymous with biodiversity conservation; it goes to the heart of remedies for the ecological crisis in their original meaning of a fight against the loss of species. This, however, is nothing really new from a storage point of view. Certainly, the idea of conservation as a practice deeply influenced by cultural values is also confirmed for biodiversity. The social construction is evident not only in the more artificial forms of conservation like seed banks or synthetic biology, but also in radical plans for protecting wild and semi-wild areas. The campaign for open access to the biodiversity heritage has highlighted the management role of public or non-profit agencies. Attempts to privatize biodiversity are at the centre of a furious legal battle that once again recalls the systemic profile of environmental issues, and the consequent necessity to promote self-establishment of forms of governance around the world (see Falkner 2003).

The active human nurturing of biodiversity is inevitable. There is a dichotomy in this process between economic exploitation and contemplative appreciation.

But conclusion is too simplistic for the social sciences. Biodiversity storage has other meanings and purposes – for example, *recreation* – that are not yet fully understood (Osti 2014) or easily dismissed (like hunting: see Fischer et al. 2012). The etymology of the word 'recreation' ('to create again, renew') touches a raw nerve in the biodiversity debate in terms of the limitation or obsolescence of the natural/artificial distinction:

> The concept of nature as external to society, either in the form of a sustenance base carrying social activity (Schnaiberg) or a sink and reservoir exploited for human progress (environmental sciences), is outdated.
> (Spaargaren, Mol and Bruynincks 2006: 18)

Storage, with its mix of old and new practices, contributes to showing the need for new distinctions (and reshufflings).

The second question that has recurred in this book is whether storage has heuristic capacity. The variety of ecological models to explain biodiversity is astonishing, and so too are attempts to translate them to human society. This quest has been pursued since the work of the Chicago School on urban ecology at the beginning of the twentieth century. Today, ecologists insist on the need for integrated models in which society and ecosystems are considered as changing together (co-evolution), albeit with different points of departure (shifting baselines). In this sense, storage is a robust pattern concerning the recurrent intersection between circulation and fixation, flow and stock, networks and storehouses (see Section 1.5). The storage/networking dichotomy works as a master frame in which to insert several human/nature combinations. The sink/source pattern applied to protected areas probably reflects this mixture: the two kinds of areas based on that pattern are kept together by an intense reciprocal flow of subpopulations.

A concept with potential heuristic value that has not emerged in this review of biodiversity storage forms is the *matrix*, defined as the 'background ecological system or land-use type in a mosaic, characterized by extensive cover, high connectivity, and/or major control over dynamics' (Forman 1995: 39). An example of a matrix is 'a forested landscape with fewer gaps in forest cover (open patches)' (Lein 2011: 37) – a biodiversity morphology that generally ensures high connectivity, and by chance resembles a typical alpine domain. These landscapes are dominated by woodland that incorporates many cleared spaces determined by human/nature combinations, like pastures, ski trails, hamlets and house clusters. In our terminology, it is a further hybrid of storage and network. Forman (1995) identifies the *mosaic* as a meta-pattern made up of patches, corridors/networks and matrixes. The challenge is to see how this metaphor of the environment is able to capture the biodiversity issue, and more precisely, which combinations are able to store more biodiversity.

The third question – whether biodiversity storage is an innovative way to address the ecological crisis – prompts a tentative but positive answer. Storage is not a radical innovation in the biodiversity field. As discussed above, it is the domain where the initial responses to the ecological crisis came about. In any case, the storage perspective makes it possible to identify a clear hierarchy of methods to maintain high biodiversity. First, extensive in situ biodiversity sanctuaries must be preserved. Then special gardens or reserves of small surface area, but with intense attention to single endangered species, are necessary. Finally, *socialized* banks can preserve some of the genetic makeup of animals and plants. However, there is no consensus on their management, so it would be wise to authorize banks only for genome conservation or selection, without inter-species mixing (i.e. GMOs). In any case, the hierarchy with in situ reserves at its top must not be inverted with the covert aim of justifying massive investments in patented nature objects. We know that it is easier for scientists to manipulate biological specimens in a controlled environment, whilst in situ research is more laborious and risks being fruitless. Location-based research and monitoring should govern nature conservation, otherwise, biodiversity storage becomes a cold archive without colours, smells and sounds – a sad prospect for something to be saved and loved.

6 Multi-storey

The fortune of the grasshoppers and the ants

In Aesop's fable, the carefree, singing grasshopper comes to appreciate the wisdom of the industrious ant, who has toiled long and hard to store food in preparation for winter, only when he finds himself dying of hunger. The fortune of the grasshoppers and the ants rests on the 'multi-storey': a play on words between the multi-level car park and the great variety of ways to store things. Many interconnected forms of storage reduce the dichotomy between savers and singers: they can live quite well, for a long time, in the same place, *when storage is manifold, widespread and modularized*. In developed countries, an extraordinary variety of storage systems exist for all kinds of goods, accompanied by high levels of waste. Austere storage practices co-exist with a profligate entertainment industry. The initial hypothesis was that people had forgotten how to save in favour of endless consumption with its related waste. But that hypothesis was soon refuted when it emerged that storage still exists. What has really changed is the mode of delivery: a formidably widespread goods distribution system has reduced any need for long-term high-capacity storage. It is also true that network distribution and the proliferation of retail outlets have led to inefficiency and waste. The provision of potable water and power in every home, as well as food shops on every street corner, have created the conditions not only for overconsumption, but also for an upsurge in many organizational fallacies. The distributive chain is full of holes, as shown by the thousands of tonnes of food squandered every day in Western countries.

The immediacy and abundance of provision has made saving and storing goods superfluous. Of course, this has happened only in one part of the world: a conspicuous minority of the world's population – in the order of a billion people – lacks these convenient but expensive distribution systems. These people are concentrated in rural areas of the global South. Nevertheless, shops and grids covering the entire territory are the model, the target, and the benchmark of all countries.

However, before being fully adopted throughout the world, this model has proved burdensome and wasteful. Old practices of storing valuable resources have been rediscovered or adapted in accordance with a new demand for eco-sustainability. *Continuity or discontinuity* with the past is therefore a good parameter on which to conduct a final evaluation of storage practices (see Table 6.1). Storage practices have not been completely abandoned in modernity. The loss of such practices is more

Table 6.1 Features of storage practices in terms of temporal continuity and module integration within networks

Module integration within networks	Continuity	Discontinuity
Low	*Biodiversity* Bio-reserves are numerous and varied; eco-corridors encounter more problems	*Water* Farm ponds have been reclaimed; new interconnected ones are only proposed
Middle/high	*Food* Widespread shops reduce household storage, but freezers and packaging increase it	*Energy* Intelligent energy grids are technically possible, but only a few experimental cases have been realized

evident for water than for food. Innovations are more marked in the energy storage sector than in biodiversity protection. The pace of change as well the periods when the need for different storage practices has become urgent vary a great deal. Renewable energy sources required a complementary storage system only a few years ago, while nature protection areas were created a hundred years ago. Nevertheless, energy storage discoveries are proceeding very rapidly, while in situ protection systems are progressing slowly and with many contradictions.

The other criterion of evaluation, as shown in Table 6.1, is *modularity*: the capacity of an organization to survive by reducing its interdependency with the environment to a minimum. This reflects how closed a system needs to be, coupled with its capacity to interact selectively with all the other systems in order to maintain *meaningfulness* (Selznick 1957). Storage capacity is precisely a way to reduce the need to continuously feed a body, a household, a settlement or a factory using external resources. At the same time, modularity indicates a system's internal articulation in order to cope with the main survival tasks. This means that storage is usually only one necessary, but minor, aspect of a more complex organization. It is coupled with production or consumption. Storage is not an alternative to the main human activities, but rather a function bestowing greater independence on a system. Overall, modularity is a specific way to integrate diversity: each module can be added or removed without altering the general order of things. Storage allows each module to be diverse, requiring few but clever adaptations to the system. Households with different levels of power self-production (and consumption styles) need access to an intelligent grid.

If a household has adequate provisions, it can survive dramatic events like drought or famine, fire or flood, disaster or earthquake. In fact, in those parts of the world that have not been subject to wars, long-term unemployment and severe climate change effects, the storage of vital resources has tended to seem pointless, tedious and strange. It has therefore been neglected. Only people living in unstable places or those who believe in an imminent catastrophe have adopted methodical storage practices. In this sense, people in poor countries exhibit the same behaviours as survivalists in rich countries, the main difference being that the former are numerous and rational, while the latter tend to be few and extravagant. Analysis of storage practices has highlighted this link between the geography of affluence and storage habits.

While storage is still routine in communities exposed to dramatic events, in the more secure parts of the world it has become a specialized activity, the prerogative of three social groups: *innovation-seekers*, *believers* in imminent radical changes, and civil protection *volunteers and officials* operating side by side. This is a diverse social panorama which traverses the traditional categories of sociology. To return to a classical interpretation, storage in Western countries is a matter of differentiation: a function delegated to a special group which is responsible for updating emergency techniques, controlling costs and educating the public. Storage, then, is in the hands of experts, usually working for large organizations like companies or professional associations (Collier and Lakoff 2008). Radical interpretations *à la* Foucault view storage practices for survival as forms of fine-tuned control over everyday habits, bodies and lives.

Such interpretations touch on an important aspect mentioned above with regard to modularity: *relative independence*. A simple but challenging example is power storage. Combined with renewable sources, this can increase a household's autonomy until it approaches nearly total self-sufficiency. With the insulation of walls and a decrease in consumption, these households can become not only passive units, requiring no energy input, but even active ones that produce a surplus of energy. This situation is challenging, because it is likely to provoke a serious crisis among the traditional energy providers. This is such a serious prospect that some countries forbid households from selling their surplus energy on the open market: they must either consume it themselves, or they must feed it into the official grid. The situation is similar for firms, which can organize massive self-production and storage of energy, but cannot sell it to other private companies. There are many intricate legal strictures, but the final outcome is clear: the energy grid seeks to remain a natural monopoly that claims the right to organize energy distribution for all.

Grid distribution certainly imposes a serious limitation on self-organized exchange, which is the full meaning of autonomy and is ideally represented by a free market. In fact, food exchange, which is less tied to a physical grid, is closer to the ideal of independence. Networks of free farmers can match the needs of likewise free consumers. Storage capacity is fundamental in achieving this freedom to sell to and buy from anybody anywhere. This is dramatically apparent in the case of seeds: if farmers cannot retain viable seeds from their own crops, they have to depend on external seed companies which will sell them products suited to their own purposes.

Storage capacity for energy as well for food is certainly an instrument of relative independence, but according to theories that are critical of claims of autonomy, seeing it as a subtle instrument of control, problems remain. Indeed, they are worsened because they are masked by noble ideals of self-reliance. Evidently, the issue is anthropological, and it concerns the ideal of humanity that every sociological theory entails (Hilgers 2011). Approaches inspired by Foucault depict an image of natural reciprocal constraint on human beings which can easily be extended to the environment.

In effect, self-storage of food, energy and water can represent false autonomy – a choice imposed from outside on households and companies. This is exemplified by the remote control of energy consumption: the capability for storage of self-produced power is rarely coupled with a complete off-grid solution. Most users prefer the partial independence provided by, for example, a combination of a PV installation, an inverter and battery storage with a grid connection to maintain a sense of greater security. This connection enables a grid manager to remotely monitor and control the household's energy consumption by means of electronic meters. Such meters are commonly used for electricity, but their use is planned for water provision as well. Thus, *remote control* – a central concept of Foucauldian approaches (Munro 2000) – finds an exemplary case. The invitation to store food, water and energy becomes the most advanced system of control over vital resources.

One way to escape this is to adopt a completely off-grid solution or to organize alternative micro-circuits of distribution, which, as mentioned above, are currently possible only for food *consumption* (for example, through farmers' markets), while the production and supply of food, water and energy are heavily constrained by the grid organization and legal norms. This is a good point for synthesis of storage practices. They are still a necessity for people living in rural areas of poor countries, but they are an ambivalent symbol of freedom where there are strictly interconnected systems of energy, water and food provision. It should be noted that biodiversity conservation and a set of storage rituals are excluded from this synthesis.

Indeed, parks and reserves, botanical gardens and seed banks provide ways to channel people's needs for access to the open air and learning activities toward arrangements imposing discipline on nature use. Many of these arrangements have a top-down origin or are dominated by experts, or by a worldwide complex of companies, foundations, governments and international NGOs – a sort of 'triple helix of nature'. Understandably, the suspicion arises that these are organizations for formatting people's subjectivity. This critical stance is welcome and useful because it helps to go beyond the banality of many interpretations of storage practices. However, a lesson from these practices is that they are irreducible to full external control because they lack a *logic*.

The most subtle of processes in modern countries is probably rationalization: Weber's *iron cage*. Storage is not free from this tendency: storage practices are developed or abandoned because they are useful or not useful to human beings. They may or may not provide a benefit, having meant survival in the past and well-being today. However, storage does not always respond to a particular objective; sometimes it is almost completely free of causes. It is irrational, extravagant and pointless.

Godelier (1996) studied gift exchange, thus helping sociology to escape from a stifling structural-functional approach. He drew arguments from the school of Marcel Mauss to demonstrate the *strength of relationships* in explaining a wide range of social processes. The role of networks in every domain of society, even those most dominated by rationality, is sufficient to highlight the merits of this school, according to which exchange is the basis of society, and before the market, reciprocity created stable bonds beyond blood and force. Godelier (1999) goes a step further by saying there are not only exchanges of equivalents, as in the market, not only free exchange of gifts as in society, but also *things that are conserved and transmitted*. Society thus consists of exchange in the three forms codified by Polanyi (1944) – market, reciprocity and redistribution – and the ceaseless storage of sacralized objects. In other words, the storage of things has a rituality that disregards functional goals and is practised for its own sake.

Godelier could not free himself from a functionalist account by saying that some objects are kept or are inalienable (or transmitted to heirs) because they represent the identity of a person or a group. Many rituals of food consumption can thus be explained: they are not functional to health or to the climate, but simply have no reason for existence other than the assertion of a distinct identity. This is plausible, and it helps to overcome many misunderstandings regarding 'extravagant' storage behaviours. We can add that storage may assume forms and meanings beyond identification with a group or a place. This is when storage becomes an organization with its own rules, as especially evident in two fields: industrial research and civil protection.

As mentioned earlier, storage is usually a phase in a more complex chain or system. It is always combined with other tasks, a minor activity compared with more 'noble' ones. In rare cases, it assumes primacy. However, a notable development in the energy sector is the formation of firms focused on the development of a sole means of storage, such as salt-based batteries. They concentrate their research on a restricted field and assume many risks of failure. The energy accumulation sector resembles the first phase of industrialization, in that many pioneers with few means but a great deal of enthusiasm are competing to develop the best-performing device – a sort of Schumpeterian world that will emerge as soon as the big energy providers have made their storage choices. Today, it is possible for utilities not to choose anything, but instead wait for external indications. Their uncertainty is similar to that of car makers, who face the same dilemmas in terms of electric-powered vehicles. Thus, the energy accumulation industry is a fluid and crowded sector with numerous companies already in crisis, but pervaded by a strong creed of storage.

The other field in which storage practices have primacy is civil protection. In this case, the storage of food, water and energy is a fundamental task that needs to be performed with care, a method to be learned through regular periods of training. It is important to conduct regular simulations of procedures and tests of the devices involved to ensure that they are still fit for purpose, and the rotation of stock is also important for similar reasons. All these actions produce a *science of civil protection* modelled on a professional association (involving affiliation, codes of conduct, unified livery and so on). Experts and trainees in storage become part of a wider world of prevention and response to disasters that is driven by the most edifying principles, and for this reason is appreciated by the public.

6.1 Policies for storage: lobbies and the need for space

The importance of storage has been recognized in every social field considered by this book. Energy and water storage have been emphasized because of the energy crisis and the effects of climate change. The importance of storing energy to compensate for intermittent renewable sources of energy is beyond doubt. Despite the general consensus, however, policies are heavily affected by resistance among the traditional energy providers, both plant and grid owners. These are often public bodies, but unfortunately, public ownership is no guarantee of innovative approaches. Thus, governments must contend with internal lobbies as well. Change can come about as a result of pressures applied by eco-consumers and segments of the industrial world involved in the production of equipment for the exploitation of renewables. At present, the latter source of pressure seems more effective, at least in those countries, like Germany and Japan, which have important interests in the battery industry. Consumers are stimulated mainly by subsidies, which public opinion has tended to regard as a waste of public money or as a hidden tax on citizens. Therefore, sustainable energy storage can arise from a holy alliance between two small minorities: utterly convinced green consumers, and the subsector of the energy accumulation industry and renewable devices producers. This tentative developing alliance is supported by governments in similarly tentative ways. The uncertainty is due to the necessity to mediate among numerous interests. A counter-example is the Californian government, whose technological neutrality makes mediation among special interests less urgent. This is probably why California has been able to undertake the boldest experiments to date in energy storage policy.

California is also a test bed for water storage policies. Its latitude and climate, together with its intensive use of water for a variety of purposes, are creating the conditions for a large-scale rationing policy unprecedented in a developed country. Rationing, usually a public measure, has a strong link with storage, as each consumer's allotted share of the resource is based on the system's capacity to accumulate water. Rationing, like compulsory reserves, is a thorny issue for decision-makers because people and companies do not like the imposition of limits on their water consumption.

The other side of water accumulation is the prevention of floods; for this reason, a new phase in the construction of artificial overflow basins has begun. Such basins are welcomed by a construction industry so severely hit by the 2008 world economic crisis. At the same time, local governments are worried about public reaction against these new reservoirs, which may occupy relatively large areas of land. More ambitious flood prevention policies need to overcome resistance from landowners, which is very difficult to achieve, especially when not coupled with material compensation for the risks involved. Nevertheless, in areas where urbanization is high and intricate – the situation in many parts of Europe – enlarging river beds is impossible, so the only solution is to instigate planned flooding areas, identified and agreed in collaboration with the entire basin population. But this means that policies must change as well: from top-down sets of rules and plans to dialogue and agreement with a wide array of stakeholders, including urban dwellers.

The traditional mechanism of incentives is urgently needed for water harvesting, because people tend not to be acutely aware of water shortages. Subsidies are necessary because some devices for recycling and accumulating water inside buildings are expensive. The social dimension emerges here in terms of shared investments in multi-apartment buildings, whose residents are usually hostile to any kind of change. The urban dimension arises as well, because dwellings of this kind are common in towns and cities. The demand for space in which to install the storage devices is high, while the roof areas of apartment blocks are small. Solar and PV panels suffer from the same problem of space.

Food is more a subjective issue, because its storage is seen as a means to save money directly or to take responsibility for managing one's own food supplies. Policies to promote demand-side food storage do not exist, probably because self-organization prevails and it is difficult to intervene in a field so jealously kept in the private sphere. Public intervention comes about for health reasons. In any case, unlike water and energy storage, there are no clear, new, more efficient tools to improve facilities in this sphere. As discussed earlier, food storage has developed over the years through some important innovations, such as canning and refrigeration, introduced many decades ago. Chemical preservatives are special cases, and their use is continuously monitored, but they are only able to provide incremental improvements in food storage. It is difficult to detect long-term effects of preservatives and packaging on health. For this reason, food authorities are usually prudent, but subject to suspicion. Storage policies are also weak on the food supply side. The policy of large-scale grain stockpiling has come to an end, and its re-adoption is only mildly debated within the Common Agriculture Policy. The calming of market prices appears justifiable to all, but its urgency depends on factors beyond production techniques, including the growing financialization of agriculture. In terms of other measures, market forces, manifested through organization such as farmers' consortiums, are invited to organize storage tools that are able to ensure the differentiation and traceability of products.

As regards the global South, the issue of supply-side food storage policies is more complicated. Most wastage occurs at the time of harvesting and storing crops, unlike in rich countries, where wastage takes place between distribution and consumption. There is a shortage of storage capacity that negatively affects local processing industries, which rely on critical masses of products. Storage is inevitably linked to more general food policies. Agriculture can no longer be conceived merely as a subsistence activity for self-sufficient villages. Consequently, vertical organization, orientation toward exports, and differentiation of products are essential for improving local food systems. Certainly, the idea of total integration in a global market, making the farm a pure centre of coordination of labour force and production means – a model pursued in Western countries – is socially and environmentally unsustainable in the South (Van der Ploeg 2008). The storage of seeds, fertilizers and crops in order to create a critical mass for middle-scale economies, possibly under the control of farmers' consortiums, is preferable. Storage capacity at local or regional level is part of a virtuous model of peasantry for the great South (Assies 2004: 123).

Storage thus works as a policy of scale: it helps to identify the best territorial scale on which to base a crop collection centre and organize around it other structures for processing and packaging food. Since dispersion of people and farms is the rule in many rural areas of poor countries, a degree of centralization of storehouses provides a compromise between the extremes of autarchic agriculture and the neoliberal global economy. This highlights an industrial model – the Modular Production Network (Sturgeon 2002) – that has recurred throughout this book and has been labelled in this concluding chapter's title as 'multi-storey': *points of accumulation work better when they are situated within a network*. Storage and networks are no longer opposing solutions. On the contrary, they can be integrated successfully in many cases: for food in low-density areas, for district heating systems coupled with household solar panels, for ponds that are interconnected in order to balance droughts and floods, not to mention biodiversity conservation, where ecological corridors among protected areas are held to be the best solution for highly urbanized areas. Policies to promote this model resemble less command and control measures and more the creation of lightly regulating agencies mediating among actors and places and able to show the advantages of the partial centralization of vital resources.

Intermediation between locations has been developed since the creation of the first botanical gardens. Given the ongoing human transformation of the environment, with the disappearance of completely untouched habitats, the compromise is to concentrate some species in certain areas and to organize continuous exchange among them. In other words, botanical gardens or animal sanctuaries become *thick nodes of a lean network*. Of course, this sort of organization is insufficient to ensure biodiversity conservation at all scales and in all forms. Consequently, nature protection policies and measures are highly articulated. An impressive variety of protected areas are planned across the world.

Biodiversity conservation is the most highly regulated field, and at the same time the domain with the greatest need for public intervention. It is necessary because biodiversity is a common good easily spoiled by private interests. Furthermore, the new frontiers of synthetic biology require the most open and public discussion about forms of conservation. Some want to recover the past, while others claim there are no limits to the creation of new living beings to be used for the conservation of biodiversity. Storage can provide an intermediate space between these two positions. The problem is also economic: discrete units of seed storage or genetically modified organisms allow rapid patent recognition. This encourages financial support for their conservation, while reducing their independence and public access.

The problem is even more acute for large protected areas that rarely have sufficient financial resources to survive. They are constantly in need of external assistance, which must inevitably derive from the exchange of species, and from financial support by visitors and donors. There is scant public money available for national parks, especially those in the poor countries – so scant, in fact, that it becomes impossible to compete with poaching and illegal logging. External funding from the richest urban areas of the world, collected through efficient networks, becomes the primary source of finance for protected areas in developing countries.

All these measures appear unable to cope with land grabbing and climate change, to mention only two global degradation trends. Geopolitical considerations are indifferent to storage needs and advantages. The rich history of storage practices is crushed under the weight of multinational companies and multi-polar politics. A specific policy on the storage of scarce resources is lacking, and it has little chance of being issued in the near future. So is this book's conclusion framed in the darkest pessimism? As touched upon at the end of each chapter, storage does not turn around the world; it is an archipelago of many niches of different scales, too weakly connected to create a critical mass for a *landscape* change (Fischer-Kowalski, Rotmans 2009). This is probably the nature of storage: nothing revolutionary or clamorous. The change provided by storage arrives in another way, despite analysts' preference for distinct radical turning points. Storage for now is an archipelago of small-scale phenomena set in a regime of continuity with the past. Nonetheless, the practices of food, water, energy and biodiversity storage show a line of progress; it is thin and curvy, but concrete and accessible to many people. This is promising in the search for a reasonable compromise between our needs for survival and the world's needs in a coming period of scarce resources.

References

Abrahamse, W., L. Steg, C. Vlek and T. Rothengatter. 2005. A review of intervention studies aimed at household energy conservation. *Journal of Environmental Psychology*, 25(3), 273–291.

ADEME. 2013. *Photovoltaic Power Applications in France in 2013*. Paris, France: ADEME for IEA PVPS.

Agar, J. 2012. Boxing clever: Heinz Wolff and the storage theory of civilisation. *UCL Science and Technology Studies Observatory*, 5 July. Available at http://blogs.ucl.ac.uk/sts-observatory/2012/07/05/boxing-clever-heinz-wolff-and-the-storage-theory-of-civilisation. Accessed 14 April 2014.

Ahvenainen, R. 2003. Active and intelligent packaging: an introduction. In *Novel Food Packaging Techniques*, edited by R. Ahvenainen. Cambridge: Woodhead Publishing, 5–21.

AIRU. 2013. *Riscaldamento Urbano – Annuario Dicembre 2013*. Milan, Italy: Associazione Italiana Riscaldamento Urbano.

Allan, J.A. 2002. *The Middle East Water Question: Hydropolitics and the Global Economy*. New York: I.B. Tauris.

Allenby, B.R. 2011. The growing gap between emerging technologies and legal-ethical oversight. In *The Growing Gap between Emerging Technologies and Legal-ethical Oversight*, edited by G.E. Marchant, B.R. Allenby and J.R. Herkert. Dordrecht, The Netherlands: Springer, 3–18.

Amilien, V. 2012. Nordic food culture – a historical perspective: interview with Henry Notaker, Norwegian culinary expert. *Anthropology of Food*. Available at http://aof.revues.org/7014. Accessed 4 February 2014.

Andersen, D.J., L. Bødker and M.V. Jensen. 2013. Large thermal energy storage at Marstal District Heating. *Proceedings of the 18th International Conference on Soil Mechanics and Geotechnical Engineering, Paris, September 2–6 2013*. London: International Society for Soil Mechanics and Geotechnical Engineering, 3,351–3,354.

Ansell, C.K. and D. Vogel, eds. 2006. *What's the Beef? The Contested Governance of European Food Safety*. Cambridge, MA: MIT Press.

Appadurai, A. 2004. The capacity to aspire: culture and the terms of recognition. In *Culture and Public Action*, edited by V. Rao and M. Walton. Palo Alto, CA: Stanford University Press, 59–84.

AsiaDHRRA. 2011. *Strengthening Household and Community Food Reserve Systems: AsiaDHRRA Experience*. http://www.iatp.org/files/451_2_107541.pdf. Accessed 5 February 2016.

References

Assies, W. 2004. From rubber estate to simple commodity production: agrarian struggles in the northern Bolivian Amazon, in *Latin American Peasants*, edited by T. Brass. London: Routledge, 83–128.

Australian Bureau of Statistics. 2010. *Environmental Issues: Water Use and Conservation.* Canberra, Australia: Australian Bureau of Statistics.

Axness, C.L. and A. Ferrando. 2011. *An Alternate Strategy for Meeting Future Urban Water Needs: Proper Water Pricing, Conservation, Rainwater Harvesting and Greywater.* Albuquerque, NM: Dept 6225, Sandia National Laboratories.

Baert, P. 1998. *Social Theory in the Twentieth Century.* Cambridge: Polity Press.

Baguma, D., W. Loiskandl and H. Jung. 2010. Water management, rainwater harvesting and predictive variables in rural households. *Water Resources Management*, 24(13), 3,333–3,348.

Baily, D. 2005. Floodplain. In *Encyclopedia of World Geography*, Vol. 1, edited by R.W. McColl New York: Infobase Publishing.

Bale, M.T. 2012. *Agriculture and Storage: How Did the First Complex Societies in Korea Develop?* Council on East Asian Studies, Yale University's Archaeology Brown Bag Lecture Series, 25 January.

Barbier, E.B. 2011. *Scarcity and Frontiers: How Economies Have Developed through Natural Resource Exploitation.* Cambridge: Cambridge University Press.

Barbour, I. 1974. *Myths, Models and Paradigms: A Comparative Study in Science and Religion.* New York: HarperCollins.

Barquín, R. 2005. The elasticity of demand for wheat in the XIV–XVIII centuries. *Revista de Historia Económica*, 23(1), 241–267.

Barraqué, B. and S. Zandaryaa. 2010. Urban water conflicts: background and conceptual framework. In *Urban Water Conflicts*, edited by B. Barraqué. Paris, France: UNESCO, 1–14.

Barraqué, B., P.S. Juuti and T.S. Katko. 2010. Urban water conflicts in recent European history: changing interactions between technology, environment and society. In *Urban Water Conflicts*, edited by B. Barraqué. Paris, France: UNESCO, 15–32.

Barrett, M. and C. Spataru. 2013. Storage in energy systems. *Energy Procedia*, 42, 670–679.

Barron, J. and J.C. Salas. 2009. *Rainwater Harvesting: A Lifeline for Human Well-being.* London: UNEP/Earthprint.

Batista, R. and M.M. Oliveira. 2009. Facts and fiction of genetically engineered food. *Trends in Biotechnology*, 27(5), 277–286.

Beamish, T.D., R. Kunkle, L. Lutzenhiser and N.W. Biggart. 2000. Why innovation happens: structured actors and emergent outcomes in the commercial buildings sector. *Consumer Behavior: 2000 American Council for an Energy-Efficient Economy.* Washington, DC: American Council for an Energy-Efficient Economy, August, 8.13–8.25.

Beardsworth, A. and T. Keil. 2002. *Sociology on the Menu: An Invitation to the Study of Food and Society.* London: Routledge.

Bellah, R.N., R. Madsen, W.M. Sullivan, A. Swidler and S.M. Tipton. 2007. *Habits of the Heart: Individualism and Commitment in American Life.* Berkeley, CA: University of California Press.

Bellassen, V. and S. Luyssaert. 2014. Carbon sequestration: Managing forests in uncertain times. *Nature*, 506, 153–155.

Bhatnagar, D., A. Currier, J. Hernandez, O. Ma and B. Kirby. 2013. *Market and Policy Barriers to Energy Storage Deployment.* Albuquerque, NM: Sandia Report SAND2013-7606.

BiodivERsA. 2014. Policy brief – nitrogen pollution and climate change reduce carbon storage and biodiversity of peatlands. *biodiversa*, 21 November. Available at http://www.biodiversa.org/693. Accessed 5 February 2016.

Black, M. and G. Strbac. 2006. Value of storage in providing balancing services for electricity generation systems with high wind penetration. *Journal of Power Sources*, 162(2), 949–953.

Blackbourn, D. 2007. *The Conquest of Nature: Water, Landscape, and the Making of Modern Germany*. New York: W.W. Norton.

Bonte, M., P.J. Stuyfzand, A. Hulsmann and P. Van Beelen. 2011. Underground thermal energy storage: environmental risks and policy developments in the Netherlands and European Union. *Ecology and Society*, 16(1), 22.

Borokini, T.I. 2013. Overcoming financial challenges in the management of botanic gardens in Nigeria: a review. *International Journal of Environmental Sciences*, 2(2), 87–94.

Borrini-Feyerabend, G., A. Kothari and G. Oviedo. 2004. *Indigenous and Local Communities and Protected Areas: Towards Equity and Enhanced Conservation*. Gland, Switzerland: IUCN, World Conservation Union.

Bowles, S. 2004. *Microeconomics: Behavior, Institutions, and Evolution*. Princeton, NJ: Princeton University Press.

Brandon, G. and A. Lewis. 1999. Reducing household energy consumption: a qualitative and quantitative field study. *Journal of Environmental Psychology*, 19(1), 75–85.

Brechin, S.R., G. Murray and C. Benjamin. 2010. Contested ground in nature protection: current challenges and opportunities in community-based natural resources and protected area management. In *Handbook of Environment and Society*, edited by J. Pretty, A. Ball, T. Benton, J. Guivant, D.R. Lee, D. Orr, M. Pfeffer and H. Ward. London: Sage, 553–577.

Brechin, S.R., P.R. Wilshusen, C.L. Fortwangler and P.C. West. 2002. Beyond the square wheel: toward a more comprehensive understanding of biodiversity conservation as social and political process. *Society and Natural Resources*, 15, 41–64.

Brookes, L. 1990. The greenhouse effect: the fallacies in the energy efficient solution. *Energy Policy*, 18, 199–201.

Brown, K. 1998. The political ecology of biodiversity, conservation and development in Nepal's Terai: confused meanings, means and ends. *Ecological Economics*, 24(1), 73–87.

Brown, L.R. 2012. *Full Planet, Empty Plates: The New Geopolitics of Food Scarcity*. New York: W.W. Norton.

Brunekreeft, G., M. Goto, R. Meyer, M. Maruyama and T. Hattori. 2014. *Unbundling of Electricity Transmission System Operators in Germany: An Experience Report*. Bremen, Germany: Jacobs University.

Bruni, A., T. Pinch and C. Schubert. 2013. Technologically dense environments: What for? What next? *TECNOSCIENZA, Italian Journal of Science and Technology Studies*, 4(2), 51–72.

Bruzzone, S. 2012. Climate change and reorganizing land use: flood control areas as a network effect. *International Journal of Urban and Regional Research*, 37(6), 2,001–2,013.

Bryant, R.L. 2002. Non-governmental organizations and governmentality: 'consuming' biodiversity and indigenous people in the Philippines. *Political Studies*, 50(2), 268–292.

Buchholz, B.M. and Z. Styczynski. 2014. *Smart Grids: Fundamentals and Technologies in Electricity Networks*. Dordrecht, The Netherlands: Springer.

Burch, D. and G. Lawrence. 2009. Towards a third food regime: behind the transformation. *Agriculture and Human Values*, 26(4), 267–279.

Buzzati, D. 1940. *Il deserto dei Tartari*. Milan, Italy: Rizzoli. Translated as *The Tartar Steppe*. Boston, MA: David R. Godine, Publisher, 2005.

Callon, M. 1987. Society in the making: the study of technology as a tool for sociological analysis. In *The social Construction of Technological Systems: New Directions in the Sociology and History of Technology*, edited by W.E. Bijker, T.P. Hughes and T.J. Pinch. Cambridge, MA: MIT Press.

Can Manufacturers Institute. 2013. *The Sustainable Solution for 21st Century Packaging*. Available at http://w.cancentral.com/sustainability/Sustainability_Report_Website.pdf. Accessed 8 May 2015.

Caprara, A., J. Wellington de Oliveira Lima, A. Correia Pequeno Marinho, P. Gondim Calvasina, L. Paes Landim and J. Sommerfeld. 2009. Irregular water supply, household usage and dengue: a bio-social study in the Brazilian northeast. *Cadernos Saúde Pública*, 25(1), S125–S136.

Caprotti, F. 2007. *Mussolini's Cities: Internal Colonialism in Italy, 1930–1939*. New York: Cambria Press.

Cardinale, B.J. 2011. Cardinale reply. *Nature*, 477, 29 September, E3–E4.

Carolan, M. 2013. *Reclaiming Food Security*. London: Routledge.

Carpanetto, D. and G. Ricuperati. 2008. *Eighteenth Century Italy. Crises, Transformations, Enlighteners*. Bari, Italy: Laterza.

Carr, E.R., P.M. Wingard, S.C. Yorty, M.C. Thompson, N.K. Jensen and J. Roberson. 2007. Applying DPSIR to sustainable development. *International Journal of Sustainable Development and World Ecology*, 14, 543–555.

Carroll, P. 2012. Water and technoscientific state formation in California. *Social Studies of Science*, 42(4), 489–516.

Carrosio, G. 2013. Energy production from biogas in the Italian countryside: policies and organizational models. *Energy Policy*, 63, 3–9.

Carrosio, G. 2015a. La valorizzazione della biodiversità nelle aree interne. Unpublished paper presented at the 'La biodiversità nascosta' conference, Rovigo, Italy, 20–21 March.

Carrosio, G. 2015b. Politiche e campi organizzativi della riqualificazione energetica degli edifici. *Sociologia Urbana e Rurale*, 37(106), 21–44.

Cartocci, R. and V. Vanelli. 2008. *Acqua, rifiuti e capitale sociale in Italia*. Bologna, Italy: Misure/Materiali di ricerca dell'Istituto Cattaneo.

Casagrande, D.G. 1997. The full circle: a historical context for urban salt marsh restoration. In *Restoration of an Urban Salt Marsh: An Interdisciplinary Approach*, edited by D. Casagrande. New Haven, CT: Yale School of Forestry and Environmental Studies, Bulletin No. 100, 13–40.

Cashdan, E.A. 1985. Coping with risk: reciprocity among the Basarwa of northern Botswana. *Man*, 20(3), 454–474.

CGIAR. 2010. *Changing Agricultural Research in a Changing World: A Strategy and Results Framework for the Reformed CGIAR*. Montpellier, France: Consultative Group on International Agricultural Research.

Chapin, M. 2004. A challenge to conservationists. can we protect natural habitats without abusing the people who live in them? *World Watch Magazine*, 17(6), 17–31.

Chen, H., T. Ngoc Cong, W. Yang, C. Tan, Y. Li and Y. Ding. 2009. Progress in electrical energy storage system: a critical review. *Progress in Natural Science*, 19(3), 291–312.

Cheng, T.C.E. and S. Podolsky. 1993. *Just-in-time Manufacturing: An Introduction*. London: Chapman and Hall.

Cherry, E. 2006. Veganism as a cultural movement: a relational approach. *Social Movement Studies*, 5(2), 155–170.

Childe, V.G. 1950. The urban revolution. *Town Planning Review*, 21(1), 3–17.

Christidis, A., C. Koch, L. Pottel and G. Tsatsaronis. 2012. The contribution of heat storage to the profitable operation of combined heat and power plants in liberalized electricity markets. *Energy*, 41(1), 75–82.

Clemons, E.K. and J.A. Santamaria. 2002. Maneuver warfare: can modern military strategy lead you to victory? *Harvard Business Review*, April, 56–65.

Codegnoni, A. 2014. Energy storage is the future of PV. *QualEnergia.it*, 29 April. Available at http://www.qualenergia.it/articoli/20140429-energy-storage-future-pv. Accessed 5 February 2016.

Coff, C. 2006. *The Taste for Ethics: An Ethic of Food Consumption*. Dordrecht, The Netherlands: Springer.

Coggins, C. 2001. Waste prevention – an issue of shared responsibility for UK producers and consumers: policy options and measurement. *Resources, Conservation and Recycling*, 32(3–4), 181–190.

Cohen, A. and L. Harris. 2014. Performing scale: watersheds as 'natural' governance units in the Canadian context. In *Performativity, Politics, and the Production of Social Space*, edited by M.R. Glass and R. Rose-Redwood. London: Routledge, 226–252.

Cohen, M.N. 1989. *Health and the Rise of Civilization*. New Haven, CT: Yale University Press.

Collier, S.J. and A. Lakoff. 2008. The vulnerability of vital systems: how 'critical infrastructure' became a security problem. In *The Politics of Securing the Homeland: Critical Infrastructure, Risk and Securitisation*, edited by M. Dunn and K.S. Kristensen. New York: Routledge.

Comfort, L.K. 2005. Risk, security, and disaster management. *Annual Review of Political Science*, 8, 335–356.

Comfort, L.K., Y. Sungu, D. Johnson and M. Dunn. 2001. Complex systems in crisis: anticipation and resilience in dynamic environments. *Journal of Contingencies and Crisis Management*, 9(3), 144–158.

Conan, M. 1999. Introduction. In *Perspectives on Garden Histories*, edited by M. Conan. Washington, DC: Dumbarton Oaks, 1–16.

Cong, W.F., J. van Ruijven, L. Mommer, G.B. De Deyn, F. Berendse and E. Hoffland. 2014. Plant species richness promotes soil carbon and nitrogen stocks in grasslands without legumes. *Journal of Ecology*, 102(5), 1,163–1,170.

Consorzio di Bonifica Adige Euganeo. 2013. *Piano di classifica per il riparto degli oneri di bonifica e di irrigazione*. Este, Italy: Consorzio di Bonifica Adige Euganeo.

Conte, E. and V. Monno. 2001. Integrating expert and common knowledge for sustainable housing management. In *Towards Sustainable Building*, edited by N. Maiellaro. Dordrecht, The Netherlands: Kluwer, 11–28.

Cordell, D., J.-O. Drangert and S. White. 2009. The story of phosphorus: global food security and food for thought. *Global Environmental Change*, 19 (2), 292–305.

Craig, R.K. 2010. Adapting water federalism to climate change impacts: energy policy, food security, and the allocation of water resources. *Environment and Energy Law and Policy Journal*, 5(2), 181–236.

Crisp, J. 2015. EU investigates government subsidies to power stations. *EurActiv* with Reuters, 29 April. Available at http://www.euractiv.com/sections/energy/eu-investigates-government-subsidies-power-stations-314201. Accessed 5 February 2016.

Crola, J.-D. 2012. *Food Reserves Must Now Be Part of International Agendas*. Paris, France: Veblen Institute for Economic Reforms.
Curtis, D.R. and M. Campopiano. 2014. Medieval land reclamation and the creation of new societies: comparing Holland and the Po Valley, 800–1500. *Journal of Historical Geography*, 44, 93–103.
Cutler, A.C. 2010. The legitimacy of private transnational governance: experts and the transnational market for force. *Socio-economic Review*, 8(1), 157–185.
Cwerner, S.B. 2001. Clothes at rest: elements for a sociology of the wardrobe. *Fashion Theory*, 5(1), 79–92.
Dahl, T.E. and G.J. Allord. 1997. *History of Wetlands in the Conterminous United States*. United States Geological Survey Water Supply Paper 2425. Available at http://water.usgs.gov/nwsum/WSP2425/history.html. Accessed 14 April 2014.
Darmon, N. and A. Drewnowski. 2008. Does social class predict diet quality? *American Journal of Clinical Nutrition*, 87(5), 1,107–1,117.
De Castro, P. Adinolfi, F. Capitanio, S. Di Falco and A. Di Mambro. 2013. *The Politics of Land and Food Scarcity*. London: Earthscan.
De Deyn, G.B., H. Quirk, Z. Yi, S. Oakley, N.J. Ostle and R.D. Bardgett. 2009. Vegetation composition promotes carbon and nitrogen storage in model grassland communities of contrasting soil fertility. *Journal of Ecology*, 97(5): 864–875.
de Moel, H., J. van Alphen, and J.C.J.H. Aerts. 2009. Flood maps in Europe – methods, availability and use. *Natural Hazards Earth System Sciences*, 9, 289–301.
De Schutter, O. 2011. *Video Message from the 17th of April for the Seed Action Days: UN Special Rapportuer [sic] on the Right to Food Speaks about the Measures Which Ensure the Monopoly of Big Seed Companies*. Available at http://www.seed-sovereignty.org/PDF/DeSchutter_Video_Text_english.pdf. Accessed 13 April 2015.
Decourt, B. and R. Debarre. 2013. *Electricity Storage*. Paris, France: Schlumberger Business Consulting Energy Institute.
Dehamna, A. 2013. Turning point for renewable energy storage. *Navigant Research Blog*, 11 November. Available at http://www.navigantresearch.com/tag/energy-storage?page=2. Accessed 13 August 2014.
Deininger, K. 1999. Making negotiated land reform work: initial experience from Colombia, Brazil and South Africa. *World Development*, 27(4), 651–672.
Denholm P. and M. Mehos. 2011. *Enabling Greater Penetration of Solar Power via the Use of CSP with Thermal Energy Storage*, NREL/TP-6A20-52978. Golden, CO: National Renewable Energy Laboratory.
DHAN Foundation. 2012. Farm ponds for enhancing food security. *Policy Brief*, 11, 1–20.
Diamond, J. 1997. *Guns, Germs and Steel: The Fates of Human Societies*. New York: W.W. Norton.
Diamond, J. 2005. *Collapse: How Societies Choose to Fail or Succeed*. London: Penguin.
Diani, M. 1995. *Green Networks: A Structural Analysis of the Italian Environmental Movement*. Edinburgh: Edinburgh University Press.
Dickens, A., C. Singh, P. Bosset, V. Mitchell, P. Cuadrado, J. Cosgrove and S. McLoughlin. 2014. *Energy Storage: Power to the People*. Available at http://www.qualenergia.it/sites/default/files/articolo-doc/Energy%20Storage.pdf. Accessed 16 September 2015.
Dickinson, F., D. Viga, I. Lizarraga and T. Castillo. 2006. Collaboration and conflict in an applied human ecology project in coastal Yucatan, Mexico. *Landscape and Urban Planning*, 74(3), 204–222.
DiMaggio, P.J. and W.W. Powell. 1991. The iron cage revisited: institutional isomorphism and collective rationality. In *The New Institutionalism in Organizational Analysis*, edited by W.W. Powell and P.J. DiMaggio. Chicago, IL: University of Chicago Press, 63–82.

References

Dinçer, İ. 2002. General introductory aspects for thermal engineering. In *Thermal Energy Storage: Systems and Applications*, edited by I. Dinçer and M. Rosen. Chichester: John Wiley, 1–55.

Directive 2007/60/EC of the European Parliament and of the Council of 23 October 2007 on the assessment and management of flood risks. *Official Journal of the European Union*, 6 November 2007, L 288/27.

Distretto Idrografico delle Alpi Orientali. 2004. *Piano di Gestione 2015–2021* (preliminary document). Venice and Trento, Italy: Direttiva Quadro Acque, 2000/60/CE.

Donati, P. 2011. *Relational Sociology: A New Paradigm for the Social Sciences*. London: Routledge.

DoE. 2013. *Grid Energy Storage*. Washington, DC: US Department of Energy.

Doom, J. 2014. SolarCity freezes energy storage program as utilities resist grid connections. *Bloomberg News*, 20 March. Available at http://www.renewableenergyworld.com/rea/news/article/2014/03/solarcity-freezes-energy-storage-program-as-utilities-resist-grid-connections. Accessed 12 August 2014.

Dore, R. 2000. *Stock Market Capitalism: Welfare Capitalism: Japan and Germany versus the Anglo-Saxons*. New York: Oxford University Press.

Douglas, E., P. Kirshen, V. Li, C. Watson and J. Wormser. 2013. *Preparing for the Rising Tide*. Boston, MA: Boston Harbor Association.

Douglas, M. 1966. *Purity and Danger*. London: Routledge and Kegan Paul.

Douthwaite, R. and G. Fallon, eds. 2010. *Fleeing Vesuvius: Overcoming the Risks of Economic and Environmental Collapse*. Gabriola Island, Canada: FEASTA, New Society Publishers.

Durkheim, E. 1912. *Les formes élémentaires de la vie religieuse*. Paris, France: Alcan.

Ecofys. 2014. *Energy Storage Opportunities and Challenges. A West Coast Perspective White Paper*. Utrecht, The Netherlands: Ecofys.

Economic Community of West African States. 2012. *Regional Food Security Reserve*. Abuja, Nigeria: Economic Community of West African States.

Edelman, B., Jaffe S. and S.D. Kominers. 2014. *To Groupon or Not to Groupon: The Profitability of Deep Discounts*. Harvard, MA: Harvard Business School, Working Paper 11-063.

Eguavoen, I. and M. McCartney. 2013. Water storage: a contribution to climate change adaptation in Africa. *Rural 21*, 47(1), 38–41.

Eisinger, P. 2002. Organizational capacity and organizational effectiveness among street-level food assistance programs. *Nonprofit and Voluntary Sector Quarterly*, 31(1), 1,115–1,130.

Ellegård, K. and B. Vilhelmson. 2004. Home as a pocket of local order: everyday activities and the friction of distance. *Geografiska Annaler*, 86(4), 281–296.

Empinotti, V.L. 2007. *Re-framing Participation: The Political Ecology of Water Management in the Lower São Francisco River Basin–Brazil*. Boulder, CO: University of Colorado, PhD thesis.

Engdahl, F.W. 2007. 'Doomsday seed vault' in the Arctic. *Global Research*, 4 December. Available at http://www.globalresearch.ca/doomsday-seed-vault-in-the-arctic-2/23503. Accessed 11 April 2015.

Environment Agency. 2009. *Achieving More: Operational Flood Storage Areas and Biodiversity*, final report. Bristol: UK Environment Agency, October.

Environmental Protection Agency. 2013. *Rainwater Harvesting Conservation, Credit, Codes, and Cost: Literature Review and Case Studies*. Washington, DC: EPA-841-R-13-002.

Eräranta, K., J. Moisander and S. Pesonen. 2009. Narratives of self and relatedness in eco-communes: resistance against normalized individualization and the nuclear family. *European Societies*, 11(3), 347–367.

Escalas, J.E. and J.R. Bettman. 2003. You are what they eat: the influence of reference groups on consumers' connections to brands. *Journal of Consumer Psychology*, 13(3), 339–348.

Escobar, A. 1998. Whose knowledge, whose nature? Biodiversity, conservation, and the political ecology of social movements. *Journal of Political Ecology*, 5(1), 53–82.

ETC Group. 2010. *Synthetic Biology: Creating Artificial Life Forms – Briefing and Recommendations for CBD Delegates to COP 10*. Ottawa, Canada: ETC Group, October.

European Commission. 2011. *The EU Cereals Regime*. Brussels, Belgium: Directorate-General for Agriculture and Rural Development, Unit C5, October.

European Union. 2012. *The Common Agricultural Policy: A Story to Be Continued*. Luxembourg: Publications Office of the European Union.

Evans A., V. Strezov and T.J. Evans. 2012. Assessment of utility energy storage options for increased renewable energy penetration. *Renewable and Sustainable Energy Reviews*, 16(6), 4,141–4,147.

Evans, P. 1989. Predatory, developmental and other state apparatuses. *Sociological Forum*, 4(4), 561–587.

Falk, B. 2013. *The Resilient Farm and Homestead: An Innovative Permaculture and Whole Systems Design Approach*. White River Junction, VT: Chelsea Green Publishing.

Falkner, R. 2003. Private environmental governance and international relations: exploring the links. *Global Environmental Politics*, 3(2), 72–87.

FAO. 2011. *The State of Food Insecurity in the World*. Rome, Italy: Food and Agriculture Organization.

Farantouris, N.E. 2009. The international and EU legal framework for the protection of wetlands with particular reference to the Mediterranean basin. *Journal of International and Comparative Environmental Law*, 6(1), 31–50.

Fargeli, R. and A.M. Wandel. 1999. Gender differences in opinions and practices with regard to a 'healthy diet'. *Appetite*, 32(2), 171–190.

Faruqui, A., S. Sergici and A. Sharif. 2010. The impact of informational feedback on energy consumption: a survey of the experimental evidence. *Energy*, 35(4), 1,598–1,608.

Feldman, D.L. 2008. Barriers to adaptive management: lessons from the Apalachicola–Chattahoochee–Flint Compact. *Society and Natural Resources*, 21(6), 512–525.

Fischer, A., V. Kereži, B. Arroyo, M. Mateos-Delibes, D. Tadie, A. Lowassa, O. Krange and K. Skogen. 2012. (De)legitimising hunting: discourses over the morality of hunting in Europe and eastern Africa. *Land Use Policy*, 32(1), 261–270.

Fischer, C. 2008. Feedback on household electricity consumption: a tool for saving energy? *Energy Efficiency*, 1(1), 79–104.

Fischer, J., B. Brosi, G.C. Daily, P.R. Ehrlich, R. Goldman, J. Goldstein, D.B. Lindenmayer, A.D. Manning, H.A. Mooney, L. Pejchar, J. Ranganathan and H. Tallis. 2008. Should agricultural policies encourage land sparing or wildlife-friendly farming? *Frontiers in Ecology and the Environment*, 6(7), 380–385.

Fischer-Kowalski, M. and J. Rotmans. 2009. Conceptualizing, observing, and influencing social–ecological transitions. *Ecology and Society*, 14(2): 3.

Fitter, R. and R. Kaplinsky. 2001. Who gains from product rents as the coffee market becomes more differentiated? A value chain analysis. *IDS Bulletin*, 32(3), 69–82.

Fleurat-Lessard, F. 2002. Qualitative reasoning and integrated management of the quality of stored grain: a promising new approach. *Journal of Stored Products Research*, 38(3), 191–218.

Forbes, S. 2008. *How Botanic Gardens Changed the World*. Adelaide, Australia: Hawke Research Institute for Sustainable Societies, University of South Australia.

Forman, R.T.T. 1995. *Land Mosaics: The Ecology of Landscapes and Regions*. Cambridge: Cambridge University Press.

Foster, B.J. 2010. Why ecological revolution? *Monthly Review*, 61(8), 1–64.

Franklin, S.H. 1969. *The European Peasantry*. London: Methuen.

Frison, C., F. Lopez and J. Esquinas-Alcazar, eds. 2012. *Plant Genetic Resources and Food Security: Stakeholder Perspectives on the International Treaty on Plant Genetic Resources for Food and Agriculture*. London: Routledge.

Fuchs, D., A. Kalfagianni and T. Havinga. 2011. Actors in private food governance: the legitimacy of retail standards and multistakeholder initiatives with civil society participation. *Agriculture and Human Values*, 28(3), 353–367.

Funtowicz, S. and J. Ravetz. 1993. Science for the post-normal age. *Futures*, 25(7), 739–755.

Gale, R.P. 1987. Resource miracles and rising expectations: a challenge to fishery managers. *Fisheries*, 12(5), 8–13.

Gallino, L. 1982. Identità, identificazione. *Laboratorio politico*, 5–6, 145–157.

Galvani, A. 2010. *I Lidi sulla costa del Delta del Po*, Comacchio, Italy: Parco del Delta del Po.

Geels, F.W. and J.W. Schot. 2007. Typology of sociotechnical transition pathways. *Research Policy*, 36(3), 399–417.

GEF Amazon Project. 2013. Get to know the strategies in place for transboundary floodable forests of the Peruvian Amazon. *Newsletter No. 3 GEF Amazon Project – Water Resources and Climate Change*, 9 November. Available at http://otca.info/gef/boletines/noticia/84. Accessed 27 January 2016.

Geman, H. and W.O. Smith. 2013. Theory of storage, inventory and volatility in the LME base metals. *Resources Policy*, 38(1), 18–28.

Gereffi, G., Humphrey, J. and T. Sturgeon. 2005. The governance of global value chains. *Review of International Political Economy*, 12(1), 78–104.

German Solar Industry Association. 2013. *Information on Support Measures for Solar Power Storage Systems*. BSW-Solar Information Paper 30, 30 August. Available at http://www.solarwirtschaft.de/fileadmin/media/pdf/infopaper_energy_storage.pdf. Accessed 5 February 2016.

Gestore Servizi Energetici. 2012. *Rapporto Statistico 2012 – Impianti a fonti rinnovabili. Settore elettrico*. Rome, Italy: Gestore Servizi Energetici.

Giblett, R.J. 1996. *Postmodern Wetlands: Culture, History, Ecology*. Edinburgh: Edinburgh University Press.

Gieryn, T.F. 1983. Boundary-work and the demarcation of science from non-science: strains and interests in professional ideologies of Scientists. *American Sociological Review*, 48(6), 781–795.

Gigerenzer, G. 1997. The modularity of social intelligence. In *Machiavellian Intelligence II: Extensions and Evaluation*, edited by A. Whiten and R.W. Byrne. Cambridge: Cambridge University Press.

Giglioli, I. and E. Swyngedouw. 2008. Let's drink to the great thirst! Water and the politics of fractured techno-natures in Sicily. *International Journal of Urban and Regional Research*, 32(2), 392–414.

Gilbert, C.L. 2011. *Food Reserves in Developing Countries: Trade Policy Options for Improved Food Security*. Geneva, Switzerland: International Centre for Trade and Sustainable Development (ICTSD), Issue Paper no. 37.

Gilroy, J.J., P. Woodcock, F.A. Edwards, C. Wheeler, C.A. Medina Uribe, T. Haugaasen and D.P. Edwards. 2014. Optimizing carbon storage and biodiversity protection in tropical agricultural landscapes. *Global Change Biology*, 20(7), 2,162–2,172.

Ginsborg, P. 2003. *A History of Contemporary Italy: Society and Politics, 1943–1988*. Basingstoke: Palgrave Macmillan.

Globevnik, L. and T. Kirn. 2009. *Small Water Bodies: Assessment of Status and Threats of Standing Small Water Bodies*. Copenhagen, Denmark: European Environmental Agency, EEA/ADS/06/001 – Water.

Godbout, J.T. and A. Caillé. 1998. *The World of the Gift*. Montreal, Canada: McGill-Queen's University Press.

Godelier, M. 1996. *L'énigme du don*. Paris, France: Fayard.

Godelier, M. 1999. Some things you give, some things you sell, but some things you must keep for yourselves: what Mauss did not say about sacred objects. In *The Enigma of Gift and Sacrifice*, edited by E. Wyschogrod, J.J. Goux and E. Boynton. New York: Fordham University Press, 19–37.

Goffman, E. 1965. Identity kits. In *Dress, Adornment and the Social Order*, edited by M.E. Roach and J.B. Eicher. New York: John Wiley, 246–264.

Goffman, E. 1983. The Interaction Order, American Sociological Association, 1982 Presidential Address. *American Sociological Review*, 48(1), 1–17.

Goldenberg, S. 2012. Drought could trigger repeat of global food crisis, experts warn. *The Guardian*, 23 July. Available at http://www.theguardian.com/environment/2012/jul/23/us-drought-global-food-crisis. Accessed 4 August 2014.

Goudsblom, J., 1995. *Fire and Civilization*. London: Penguin.

Gramsci, A. 2014. *Quaderni dal carcere*. Turin, Italy: Einaudi, first published 1948–51.

Granovetter, M.S. 1973. The strength of weak ties. *American Journal of Sociology*, 78(6), 1,360–1,380.

Green, A. 2013. *Education and State Formation: Europe, East Asia and the USA*. Basingstoke: Palgrave Macmillan.

Gregoire, C. 2014. How awe-inspiring experiences can make you happier, less stressed and more creative. *The Huffington Post*, 22 September. Available at http://www.huffingtonpost.com/2014/09/22/the-psychology-of-awe_n_5799850.html. Accessed 8 August 2015.

Grizzetti, B. 2011. Nitrogen as a threat to European water quality. In *The European Nitrogen Assessment: Sources, Effects and Policies Perspectives*, edited by M.A. Sutton, C.M. Howard, J.W. Erisman, G. Billen, A. Bleeker, P. Grennfelt, H. van Grinsven and B. Grizzetti. Cambridge: Cambridge University Press, 379–404.

Groupe de Bruges. 2012. *A CAP for the Future!?* Wageningen, The Netherlands: Groupe de Bruges.

Guastoni, C. 2004. *L'acquedotto civico di Genova: un percorso al futuro*. Milan, Italy: FrancoAngeli.

Gustafson, A., S. Fleischer and A. Joelsson. 2000. A catchment-oriented and cost-effective policy for water protection. *Ecological Engineering*, 14(4), 419–427.

Haering, S.A. and S.B. Syed. 2009. *Community Food Security in United States Cities: A Survey of the Relevant Scientific Literature*. Baltimore, MD: John Hopkins Center for a Livable Future.

Halewood, M., I. López Noriega and S. Louafi, eds. 2013. *Crop Genetic Resources as a Global Commons: Challenges in International Law and Governance*. London: Routledge.

Halkier, B. 2009. A practice theoretical perspective on everyday dealings with environmental challenges of food consumption. *Anthropology of Food*. Available at http://aof.revues.org/6405. Accessed 24 December 2013.

Hall, N.D. 2006. Toward a new horizontal federalism: interstate water management in the Great Lakes region. *Colorado Law Review*, 77(2), 405–456.

Hall, P.A. and R.C.R Taylor. 1996. Political science and the three 'new' institutionalisms. *Political Studies*, 44(5), 936–957.

Halstead, P. and J.M. O'Shea. 1982. A friend in need is a friend indeed: social storage and the origins of social ranking. In *Ranking, Resource and Exchange*, edited by C. Renfrew. Cambridge: Cambridge University Press, 92–99.

Hand, M. and E. Shove. 2007. Condensing practices: ways of living with a freezer. *Journal of Consumer Culture*, 7(1), 79–104.

Hannigan, J. 2006. *Environmental Sociology*. London: Routledge.

Hansen, A.J. 2011. Contribution of source–sink theory to protected area science. In *Sources, Sinks, and Sustainability across Landscapes*, edited by J. Liu, V. Hull, A. Morzillo and J. Wiens. Cambridge: Cambridge University Press, 339–360.

Haq, G., H. Cambridge and A. Owen. 2013. A targeted social marketing approach for community pro-environmental behavioural change. *Local Environment*, 18(10), 1,134–1,152.

Hargreaves, T. 2011. Practice-ing behaviour change: applying social practice theory to proenvironmental behaviour change. *Journal of Consumer Culture*, 11(1), 79–99.

Hartmann, T. 2011. *Clumsy Floodplains: Responsive Land Policy for Extreme Floods*. Farnham: Ashgate.

Hasnain, S.M. 1998. Review on sustainable thermal energy storage technologies, Part I: heat storage materials and techniques. *Energy Conversion and Management*, 39(11), 1,127–1,138.

Hauber J. and C. Ruppert-Winkel. 2012. Moving towards energy self-sufficiency based on renewables: comparative case studies on the emergence of regional processes of socio-technical change in Germany. *Sustainability*, 4, 491–530.

Hellberg, S. 2014. Water, life and politics: exploring the contested case of eThekwini municipality through a governmentality lens. *Geoforum*, 56, 226–236.

Hendon, J.A. 2000. Having and holding: storage, memory, knowledge, and social relations. *American Anthropologist*, 102(1), 42–53.

Hendriks, M., A. Snijders and N. Boid. 2008. *Underground thermal energy storage for efficient heating and cooling of buildings*. London: IfTech Publications.

Héritier, S. 2010. Participation et gestion dans les parcs nationaux de montagne: approches anglo-saxonnes. *Revue de Géographie Alpine*, 98–1, 29 March. Available at http://rga.revues.org/1128. Accessed 5 February 2016.

Hernández-Morcillo, M., J. Hoberg, E. Oteros-Rozas, T. Plieninger, E. Gómez-Baggethun and V. Reyes-García. 2014. Traditional ecological knowledge in Europe: status quo and insights for the environmental policy agenda. *Environment*, 56(1), 3–17.

Hicks, C., S. Woroniecki, M. Fancourt, M. Bieri, H. Garcia Robles, K. Trumper and R. Mant. 2014. *The Relationship between Biodiversity, Carbon Storage and the Provision of Other Ecosystem Services: Critical Review for the Forestry Component of the International Climate Fund*. Cambridge: UNEP-WCMC.

Higman, B.W. 2011. *How Food Made History*. Chichester: Wiley-Blackwell.

Hilgers, M. 2011. *The Three Anthropological Approaches to Neoliberalism*. London: Blackwell, 351–363.

Hinrichs, C.C. 2013. Regionalizing food security? Imperatives, intersections and contestations in a post-9/11 world. *Journal of Rural Studies*, 29(4), 7–18.

Hodgson, J.A., W.E. Kunin, C.D. Thomas, T.G. Benton and D. Gabriel. 2010. Comparing organic farming and land sparing: optimizing yield and butterfly populations at a landscape scale. *Ecology Letters*, 13(11), 1,358–1,367.

Horton, R. 1960. A definition of religion, and its uses. *Journal of the Royal Anthropological Institute*, 90, 201–226.

Huang, S.L., Y.C. Lee, W.W. Budd and M.C. Yang. 2012. Analysis of changes in farm pond network connectivity in the peri-urban landscape of the Taoyuan area, Taiwan. *Environmental Management*, 49(4), 915–28.

Humphery, K. 2009. *Excess: Anti-consumerism in the West.* Cambridge: Polity Press.

IEA-ETSAP and IRENA. 2013. *Thermal Energy Storage Technology Brief.* Abu Dhabi, United Arab Emirates: IEA-ETSAP and IRENA, Technology Brief E17.

International Civil Society Working Group on Synthetic Biology. 2011. *A Submission to the Convention on Biological Diversity's Subsidiary Body on Scientific, Technical and Technological Advice (SBSTTA) on the Potential Impacts of Synthetic Biology on the Conservation and Sustainable Use of Biodiversity*, 17 October. Available at https://www.cbd.int/doc/emerging-issues/Int-Civil-Soc-WG-Synthetic-Biology-2011-013-en.pdf. Accessed 5 February 2016.

International Food Information Council Foundation. 2010. *Understanding Our Food Communications Tool Kit.* Washington, DC: International Food Information Council Foundation.

International Water Management Institute. 2009. *Flexible Water Storage Options: For Adaptation to Climate Change.* Colombo, Sri Lanka: International Water Management Institute, Water Policy Brief 31.

Ipsos MORI. 2011. Global citizen reaction to the Fukushima nuclear plant disaster. *Ipsos MORI*, 23 June. Available at https://www.ipsos-mori.com/researchpublications/researcharchive/2817/Strong-global-opposition-towards-nuclear-power.aspx. Accessed 21 September 2015.

ISMEA. 2014. *Censimento dei centri di stoccaggio dei cereali.* Rome, Italy: Ministero delle Politiche Agricole, Alimentari e Forestali.

Istat. 2012. *Giornata Mondiale dell'Acqua. Le statistiche dell'Istat.* Rome, Italy: Istat.

Jackson, T. 2009. *Prosperity without Growth: Economics for a Finite Planet.* London: Earthscan.

Jager, H.I., R.A. Efroymson, J.J. Opperman and M.R. Kelly. 2015. Spatial design principles for sustainable hydropower development in river basins. *Renewable and Sustainable Energy Reviews*, 45, 808–816.

Jasanoff, S. 1987. Contested boundaries in policy-relevant science. *Social Studies of Science*, 17(2), 195–230.

Jedlowski, P. 2009. *Il mondo in questione. Introduzione alla storia del pensiero sociologico.* Rome, Italy: Carocci.

Jefferson, M. 1939. The law of the primate city. *Geographical Review*, 29, 226–232.

Jones, G.E. and C. Garforth. 1997. The history, development, and future of agricultural extension. In *Improving Agricultural Extension: A Reference Manual*, edited by B.E. Swanson, R.P. Bentz and A.J. Sofranko. Rome, Italy: Food and Agriculture Organization, 3–12.

Jones, P. and N. Macdonald. 2007. Making space for unruly water: Sustainable drainage systems and the disciplining of surface runoff. *Geoforum*, 38, 534–544.

Juntunen, J.K. 2014. *Prosuming Energy: User Innovation and New Energy Communities in Renewable Micro-generation*. Aalto, Finland: Aalto University, PhD dissertation.

Kakade, B.K., G.S. Neelam, K.J. Petare and C. Doreswamy. 2002. *Rejuvenation of Rivulets: Farm Pond Based Watershed Development*. Pune, India: BAIF Development Research Foundation.

Kasperson, R.E., O. Renn, P. Slovic, H.S. Brown, J. Emel, R. Goble, J.X. Kasperson and S. Ratick. 1988. The social amplification of risk: a conceptual framework. *Risk Analysis*, 8(2), 177–187.

Kaufman, S., 2006. Symbolic politics or rational choice? Testing theories of extreme ethnic violence. *International Security*, 30(4), 45–86.

Keynes, J.M. 1938. The policy of government storage of food-stuffs and raw materials. *Economic Journal*, 48(191), 449–460.

Kintisch, E. 2015. Amazon rainforest ability to soak up carbon dioxide is falling, *Science*, 18 March. Available at http://www.sciencemag.org/news/2015/03/amazon-rainforest-ability-soak-carbon-dioxide-falling. Accessed 15 April 2015.

Kloppenburg, J., Jr. 2010. Seed sovereignty: the promise of open source biology. In *Food Sovereignty: Reconnecting Food, Nature and Community*, edited by A. Desmarais, H.K. Wittman and N. Wiebe. Black Point, Canada: Fernwood Publishing, 152–167.

Kloss, C. 2008. *Rainwater Harvesting Policies, Municipal Handbook*. Washington, DC: Environmental Protection Agency, EPA-833-F-08–010.

Knox-Hayes, J., M.A. Brown, B.K. Sovacool and Y. Wang. 2013. *Understanding Attitudes toward Energy Security: Results of a Cross-national Survey*. Atlanta, GA: School of Public Policy, Georgia Institute of Technology, Working Paper 74.

Krusemark, K. and C. Block. 2011. Historical and contemporary cases illustrating the vulnerability of specific commodities and sectors. In *Food and Agriculture Security: An Historical, Multidisciplinary Approach*, edited by J. Kastner. Santa Barbara, CA: ABC-CLIO, 61–82.

Krystallis, A., L. Frewer, G. Rowe, J. Houghton, O. Kehagia and T. Perrea. 2007. A perceptual divide? Consumer and expert attitudes to food risk management in Europe. *Health, Risk and Society*, 9(4), 407–424.

Lackey, R.T. 2003. Appropriate use of ecosystem health and normative science in ecological policy. In *Managing for Healthy Ecosystems*, edited by D.J. Rapport, William L. Lasley, D.E. Rolston, N.O. Nielsen, C.O. Qualset and A.B. Damania. Boca Raton, FL: Lewis Publishers, 175–186.

Lamy, P. 2013. *Millennium Rage: Survivalists, White Supremacists, and the Doomsday Prophecy*. New York: Plenum Press.

Laudiero, S. 2012. *Ti presento i preppers*. Turin, Italy: Zandegù.

Lehmann, P., F. Creutzig, M.-H. Ehlers, N. Friedrichsen, C. Heuson, L. Hirth and R. Pietzcker. 2012. Carbon lock-out: advancing renewable energy policy in Europe. *Energies*, 5(2), 323–354.

Lein, J.K. 2011. *Environmental Sensing: Analytical Techniques for Earth Observation*. New York: Springer.

León, D., P. Peñalver, J. Casas, M. Juan, F. Fuentes, I. Gallego and J. Toja. 2010. Zooplankton richness in farm ponds of Andalusia (southern Spain): a comparison with natural wetlands. *Limnetica*, 29(1), 153–162.

Lévi-Strauss, C. 1970. *The Raw and the Cooked*. London: Jonathan Cape.

Lewis, R. 1952. *Edwin Chadwick and the Public Health Movement 1832–1854*. London: Longman.

Liao, K. 2012. A theory on urban resilience to floods: a basis for alternative planning practices. *Ecology and Society*, 17(4), 48.

Lighart, T.N., A.M.M. Ansems and J. Jetten. 2005. *Eco-efficiency and Nutritional Aspects of Different Product-packaging Systems: An Integrated Approach towards Sustainability.* Available at http://www.apeal.org/wp-content/uploads/2015/03/22-11-2005_TNO-Study-exec-summary.pdf. Accessed 5 February 2016.

Linskey, A. 2013. Greenwich stilt houses foreshadow impact of new FEMA maps. *BloombergBusiness*, 19 August. Available at http://www.bloomberg.com/news/2013-08-19/greenwich-stilt-houses-foreshadow-impact-of-new-fema-maps.html. Accessed 21 September 2015.

Lockwood, M., C. Kuzemko, C. Mitchell and R. Hoggett. 2013. *Theorising Governance and Innovation in Sustainable Energy Transitions.* Exeter: University of Exeter, EPG Working Paper 1304.

Lombardi, P. 2015. The analytic network process method for territorial integrated evaluation. In *Smart Evaluation and Integrated Design in Regional Development*, edited by G. Brunetta. Farnham: Ashgate, 75–96.

Löwith, K. 1953. *Weltgeschichte und Heilsgeschehen.* Stuttgart, Germany: Kohlhammer.

Luo, X., J. Wang, M. Dooner and J. Clarke. 2015. Overview of current development in electrical energy storage technologies and the application potential in power system operation. *Applied Energy*, 137(1), 511–536.

Lyons, C. 2013. *Grid-scale Energy Storage in North America 2013: Applications, Technologies and Suppliers.* Boston, MA: GTM Research.

MacCoun, R.J. 1998. Biases in the interpretation and use of research results. *Annual Review of Psychology*, 49, 259–287.

Macintyre, S., A. Ellaway and S. Cummins. 2002. Place effects on health: how can we conceptualise, operationalise and measure them? *Social Science and Medicine*, 55(1), 125–139.

MacKerron, C.B. 2015. *Waste and Opportunity 2015: Environmental Progress and Challenges in Food, Beverage, and Consumer Goods Packaging.* New York: Natural Resources Defense Council and As You Sow.

MacKinnon D. 2000. Managerialism, governmentality and the state: a neo-Foucauldian approach to local economic governance. *Political Geography*, 19(3), 293–314.

Maestrelli, A. and M. Della Campa. 2011. Work in progress: miniaturization of food processing equipment for small enterprises and use of renewable energy. Paper presented at CIGR Section VI International Symposium, Nantes, France, 18–20 April.

Mago, P.J., R. Luck and A. Knizley. 2014. Combined heat and power systems with dual power generation units and thermal storage. *International Journal of Energy Research*, 38(7), 896–907.

Malashenko, E., R. Lee, C. Villarreal, T. Howard and A. Gupta. 2012. *CPUC Energy Storage Proceeding R.10-12-007: Energy Storage Framework Staff Proposal.* San Francisco, CA: California Public Utilities Commission, 3 April.

Manders, T. 2011. ed. *Scarcity in a Sea of Plenty? Global Resource Scarcities and Policies in the European Union and the Netherlands.* The Hague, The Netherlands: PBL Netherlands Environmental Assessment Agency.

Manuta, J., S. Khrutmuang, D. Huaisai and L. Lebel. 2006. Institutionalized incapacities and practice in flood disaster management in Thailand. *Science and Culture*, 72(1–2): 10–22.

Marchetti, M. 2002. Environmental changes in the central Po plain (northern Italy) due to fluvial modifications and anthropogenic activities. *Geomorphology*, 44(3–4), 361–373.

Markham, W.T. 2008. *Environmental Organisations in Modern Germany: Hardy Survivors in the Twentieth Century and Beyond.* New York: Berghahn Books.

Markusson, N., S. Shackley and E. Benjamin, eds. 2012. *The Social Dynamics of Carbon Capture and Storage: Understanding CCS Representations, Governance and Innovation*. London: Routledge.

Marquardt, J. 2014. Energy transition in Southeast Asia: how multi-level governance systems affect renewable energy projects. Paper presented at ECPR Graduate Student Conference, Innsbrück, Germany, 3–5 July.

Martin, P., A. Newton and J. Bullock. 2013. Carbon pools recover more quickly than plant biodiversity in tropical secondary forests. *Proceedings of the Royal Society B*, 281(1,782), 6 November, corrected 7 May 2014. Available at http://rspb.royalsocietypublishing.org/content/280/1773/20132236. Accessed 5 February 2016.

Martínez Alier, J. 2009. *Ecologia dei poveri. La lotta per la giustizia ambientale*. Milan, Italy: Jaca Books. Originally published in Spanish, 2004.

Martinot, E. 2015. How is Germany integrating and balancing renewable energy today? *Energy Transition: The German Energiewende*, 19 February. Available at http://energytransition.de/2015/02/how-germany-integrates-renewable-energy/. Accessed 16 June 2015.

Massarutto, A. and A. de Carli. 2014. Two birds with one stone: improving ecological quality and flood protection through river restoration in Northern Italy. *Economics and Policy of Energy and the Environment*, 1, 93–121.

Maxim, L., Spangenberg J.H. and M. O'Connor. 2009. An analysis of risks for biodiversity under the DPSIR framework. Ecological Economics, 69 (1), 12–23.

Maxwell, S. 1996. Food security: a post-modern perspective. *Food Policy*, 21(2), 155–170.

Mays, L., G.P. Antoniou and A.N. Angelakis. 2013. History of water cisterns: legacies and lessons. *Water*, 5(4), 1,916–1,940.

Mazur, A. 2013. Energy and electricity in industrial nations: the sociology and technology of energy, London: Earthscan.

McEvoy, J. 2014. Desalination and water security: the promise and perils of a technological fix to the water crisis in Baja California Sur, Mexico. *Water Alternatives*, 7(3), 518–541.

McKeen, L.W. 2012. *The Effect of Sterilization on Plastics and Elastomers*. Oxford: William Andrew.

McLarnon, F.R. and E.J. Cairns. 1989. Energy storage. *Annual Review of Energy*, 14, 241–271.

McMichael, A.J., J.W. Powles, C.D. Butler and R. Uauy. 2007. Food, livestock production, energy, climate change, and health. *The Lancet*, 370(9,594), 1,253–1,263.

McMichael, P. 2009. A food regime analysis of the 'world food crisis'. *Agriculture and Human Values*, 26(4): 281–295.

Meneghello, G. 2012. Ecco perché l'Italia deve credere alla geotermia a bassa entalpia. *QualEnergia.it*, 6 December. Available at http://www.qualenergia.it/articoli/20121206-italia-deve-credere-a-geotermia-a-bassa-entalpia-pompe-calore-geotermiche. Accessed 5 February 2016.

Meneghello, G. 2013. Cosa c'è dietro al nuovo divieto di accumuli del GSE? *QualEnergia.it*, 23 September. Available at http://www.qualenergia.it/articoli/20130923-cosa-c-%C3%A8-dietro-al-nuovo-divieto-di-accumuli-del-gse. Accessed 5 February 2016.

Mercalli, L. 2012. *Prepariamoci*. Florence, Italy: Chiarelettere.

Meyer, H. and W. Hermans. 2009. Adaptive strategies and the Rotterdam floodplain, Delft, The Netherlands: TU-Delft.

Michels, R. 1962. *Political Parties: A Sociological Study of the Oligarchical Tendencies of Modern Democracy*. New York: Collier Books; first published 1911.

Midttømme, K., D. Banks, R.K. Ramstad, O.M. Sæther and H. Skarphagen. 2008. Ground-source heat pumps and underground thermal energy storage: energy for the future. In *Geology for Society*, edited by T. Slagstad. Trondheim, Norway: Norges Geologiske Undersøkelse, 93–98.

Miller, J.W. 2009. *Farm Ponds for Water, Fish and Livelihoods*. Rome, Italy: Food and Agriculture Organization, Diversification Booklet no. 13.

Millstone, E. 2009. Science, risk and governance: radical rhetorics and the realities of reform in food safety governance. *Research Policy*, 38, 624–636.

Mische, A. 2011. Relational sociology, culture, and agency. In *The Sage Handbook of Social Network Analysis*, edited by J. Scott and P. Carrington, London: Sage, 80–97.

Mishra, A., P. Prabuthas, and H.N. Mishra. 2012. Grain storage: methods and measurements. *Quality Assurance and Safety of Crops and Foods*, 4(3), 144.

Mitchell, C. 2007. *The Political Economy of Sustainable Energy*. Basingstoke: Palgrave.

Moisander, J. and S. Pesonen. 2002. Narratives of sustainable ways of living: constructing the self and the other as a green consumer. *Management Decision*, 40(4), 329–342.

Molle, F., P.P. Mollinga and R. Meinzen-Dick. 2008. Water, politics and development: introducing water alternatives. *Water Alternatives*, 1(1), 1–6.

Mollinga, P.P. 2008. Water, politics and development: framing a political sociology of water resources management. *Water Alternatives*, 1(1), 7–23.

Monti, A. 1998. *I braccianti. L'epica dell'Italia contadina*. Bologna, Italy: Il mulino.

Mori, P.A. 2013. Customer ownership of public utilities: new wine in old bottles. *Journal of Entrepreneurial and Organizational Diversity*, 2(1), 54–74.

Moritz, M. 2013. Traditional economies. In *International Encyclopedia of Economic Sociology*, edited by J. Beckert and M. Zafirovski. London: Routledge, 677–679.

Morton, L.W. and K.R. Olson. 2013. Birds Point–New Madrid Floodway: redesign, reconstruction and restoration. *Journal of Soil and Water Conservation*, 68(2), 35A–40A.

Morton, L.W. and C.Y. Weng. 2009. Getting to better water quality outcomes: the promise and challenge of the citizen effect. *Agriculture and Human Values*, 26(1–2), 83–94.

Mosselmans, B. 2013. Scarcity. In *International Encyclopedia of Economic Sociology*, edited by J. Beckert and M. Zafirovski. London: Routledge, 590–591.

Mostert, E. and S.J. Junier. 2009. The European Flood Risk Directive: challenges for research. *Hydrology and Earth System Sciences Discussions*, 6(4), 4,961–4,988.

Moyo, S. and P. Yeros, eds. 2005. *Reclaiming the Land: The Resurgence of Rural Movements in Africa, Asia and Latin America*. London: Zed Books.

Munro, L. 2000. Non-disciplinary power and the network society. *Organization*, 7(4), 679–695.

Nally, D. 2011. The biopolitics of food provisioning. *Transactions of the Institute of British Geographers*, 36, 37–53.

Natali, F., A. Dalla Marta, F. Orlando and S. Orlandini. 2009. Water use in Italian agriculture: analysis of rainfall patterns, irrigation systems and water storage capacity of farm ponds. *Italian Journal of Agrometeorology*, 55(3), 56–59.

National Assembly of Wales. 2013. *EU Policy update (EU2013.04): Reform of Common Agricultural Policy (CAP),*. Cardiff: National Assembly of Wales Research Service, EU Policy, 1–5.

Navarro, P. 1996. The Japanese electric utility industry. In *International Comparison of Electricity Regulation*, edited by R.J. Gilbert and E.P. Kahn. Cambridge: Cambridge University Press, 231–276.

Ness, B., S. Anderberg and L. Olsson. 2010. Structuring problems in sustainability science: the multi-level DPSIR framework. *Geoforum*, 41(3), 479–488.

Newman, M.E.J. 2006, Modularity and community structure in networks, *Proceedings of the National Academy of Sciences of the USA*, 103(23), 8,577–8,582.
Nordström, K., C. Coff, H. Jönsson, L. Nordenfelt and U. Görman. 2013. Food and health: individual, cultural, or scientific matters? *Genes and Nutrition*, 8(4), 357–363.
Norman, E.S., C. Cook and A. Cohen. 2015. Introduction: why the politics of scale matter in the governance of water. In *Negotiating Water Governance*, edited by E.S. Norman, C. Cook and A. Cohen. Farnham: Ashgate, 1–25.
Nuytten, T., B. Claessens, K. Paredis, J. Van Bael and D. Six. 2013. Flexibility of a combined heat and power system with thermal energy storage for district heating. *Applied Energy*, 104, 583–591.
Ó Gráda, C. and J.-M. Chevet. 2002. Famine and market in ancien régime France. *Journal of Economic History*, 62(3), 706–733.
OECD/IEA. 2014. *Technology Roadmap: Energy Storage*. Paris, France: IEA Publishing.
Onaindia, M., B. Fernández de Manuel, I. Madariaga and G. Rodríguez-Loinaz. 2013. Co-benefits and trade-offs between biodiversity, carbon storage and water flow regulation. *Forest Ecology and Management*, 289, 1–9.
Orlandi, E.C. 2015, Oltre lo sporco: uno studio etnografico sulla classificazione degli scarti in un ipermercato coop. *Rassegna Italiana di Sociologia*, 2, LVI(2), 179–204.
Osti, G. 2006. *Nuovi Asceti. Consumatori, imprese e istituzioni di fronte alla crisi ambientale*. Bologna, Italy: Il mulino.
Osti, G. 2012a. Frames, organisations, and practices as social components of energy. *International Review of Sociology*, 22(3), 412–428.
Osti, G. 2012b. Wind energy exchanges and rural development in Italy. In *Sustainability and Short-term Policies: Improving Governance in Spatial Policy Interventions*, edited by S. Sjöblom, K. Andersson, T. Marsden and S. Skerratt. Farnham: Ashgate, 245–259.
Osti, G. 2014. Agroecologia e *buen vivir*. Come far giocare l'uomo e l'ambiente. In *Le sfide della sostenibilità. Il* buen vivir *andino dalla prospettiva europea*, edited by S. Baldin and M. Zago. Bologna, Italy: Filodiritto, 253–268.
Ostrom, V. and E. Ostrom. 1977. Public goods and public choices. In *Alternatives for Delivering Public Services: Towards Improved Performance*, edited by E.S. Savas. Boulder, CO: Westview Press, 7–49.
Owen, B. 2008. *Some Theories of the Origins of Civilization – Batch 1*. Available at http://bruceowen.com/emciv/341-08f-09-Theories.pdf. Accessed 23 March 2014.
Pacyga, D. 2008. Chicago: slaughterhouse to the world. In *Meat, Modernity and the Rise of the Slaughterhouse*, edited by P. Young Lee. Durham, NH: University of New Hampshire Press, 153–166.
Paek, H.J., K. Hilyard, V. Freimuth, J.K. Barge and M. Mindlin. 2010. Theory-based approaches to understanding public emergency preparedness: implications for effective health and risk communication. *Journal of Health Communication*, 15(4), 428–44.
Paksoy, H. 2013. Thermal energy storage today. Paper presented at IEA Energy Storage Technology Roadmap Stakeholder Engagement Workshop, Paris, France, 14 February.
Palermo, G. 2003. *Equilibrio economico generale e fallimenti del mercato, Teoria, metodologia e filosofia morale*. Brescia, Italy: CLUB.
Parra, D., M. Gillott, S.A. Norman and G.S. Walker. 2015. Optimum community energy storage system for PV energy time-shift. *Applied Energy*, 137, 576–587.
Paul, K.T. 2008. *Thought for Food (Safety) in the EU: A Discourse-analytical Approach*. Amsterdam, The Netherlands: University of Amsterdam, GARNET Working Paper no. 38/08.

Pauschinger, T. 2011. *Supplying Renewable Zero-emission Heat*. Available at http://solar-district-heating.eu/Portals/8/%C3%96sterreich%20Uploads/Brochure_EN.pdf. Accessed 21 April 2014.
Paxton, R.O. 1997. *French Peasant Fascism: Henry Dorgeres' Greenshirts and the Crises of French Agriculture, 1929–1939*. Oxford: Oxford University Press.
Pellizzoni, L. 2010. Environmental knowledge and deliberative democracy. In *Environmental Sociology: European Perspectives and Interdisciplinary Challenges*, edited by M. Gross and H. Heinrichs. Berlin, Germany: Springer, 159–182.
Pellizzoni, L. 2011a. Governing through disorder: neoliberal environmental governance and social theory. *Global Environmental Change*, 21(3), 795–803.
Pellizzoni, L. 2011b. The politics of facts: local environmental conflicts and expertise. *Environmental Politics*, 20(6), 765–785.
Pellizzoni, L. 2015. *Ontological Politics in a Disposable World: The New Mastery of Nature*. Farnham: Ashgate.
Perna, T. 2011. Eventi estremi. Come salvare il pianeta e noi stessi dalle tempeste climatiche e finanziarie. Milan, Italy: Altraeconomia.
Perren, R. 1990. Structural change and market growth in the food industry: flour milling in Britain, Europe, and America, 1850–1914. *Economic History Review*, 43(3), 420–437.
Petrash, J.M. 2006. Long-term natural gas contracts: dead, dying, or merely resting? *Energy Law Journal*, 27(2), 545–582.
Pettenger, M.E. 2007. *The Social Construction of Climate Change: Power, Knowledge, Norms, Discourses*. Farnham: Ashgate.
Piccioni, L. 2013. The Abruzzo National Park and nature protection in Italy: the recurrence of a centrality. In *Ninety Years of the Abruzzo National Park 1922–2012: Proceedings of the Conference Held in Pescasseroli, May 18–20, 2012*, edited by L. Piccioni. Newcastle upon Tyne: Cambridge Scholars Publishing, 103–114.
Pigni, M. 2015. L'accumulo di energia elettrica ai blocchi di partenza. *QualEnergia.it*, 10 April. Available at http://www.qualenergia.it/articoli/20150410-accumulo-di-energia-elettrica-ai-blocchi-di-partenza-storage. Accessed 5 February 2016.
Pimental, D. and M. Pimental. 2008. Food processing, packaging, and preparation. In *Food, Energy, and Society*, 3rd edn, edited by D. Pimental and M. Pimental. Boca Raton, FL: CRC Press.
Podgornik, A., B. Sucic, D. Stanicic and P. Bevk. 2013. The impact of smart metering on energy efficiency in low-income housing in Mediterranean. In *Climate-smart Technologies*, edited by W. Leal Filho, F. Mannke and R. Mohee. Berlin, Germany: SpringerLink, 597–614.
Polanyi, K. 1944. *The Great Transformation: The Political and Economic Origins of Our Time*. Boston, MA: Beacon Press.
Poli, R. 2010. An introduction to the ontology of anticipation. *Futures: The Journal of Policy, Planning and Future Studies*, 42(7), 769–776.
Pollitt, M. 2007. The Arguments For and Against Ownership Unbundling of Energy Transmission Networks. Cambridge Working Papers in Economics. Available at https://www.repository.cam.ac.uk/handle/1810/194717. Accessed 5 February 2016.
Potter, N.N. and J.H. Hotchkiss. 1998. *Food Science*. New York: Springer.
Pringle, C.M., M.C. Freeman and B.J. Freeman. 2000. Regional effects of hydrologic alterations on riverine macrobiota in the New World: tropical–temperate comparisons. *BioScience*, 50(9), 807–823.
Punch, S., N. Dorrer, R. Emond and I. McIntosh. 2009. *Food Practices in Residential Children's Homes: The Views and Experiences of Staff and Children*. Stirling: Department of Applied Social Science, University of Stirling.

Ragheb, M. 2014. *Solar Thermal Power and Energy Storage: Historical Perspective.* Available at http://www.solarthermalworld.org/sites/gstec/files/story/2015-04-18/solar_thermal_power_and_energy_storage_historical_perspective.pdf. Accessed 5 February 2016.

Ranganathan, M. 2014. 'Mafias' in the waterscape: urban informality and everyday public authority in Bangalore. *Water Alternatives*, 7(1): 89–105.

Rashid, S. and S. Lemma. 2011. *Strategic Grain Reserves in Ethiopia. Institutional Design and Operational Performance.* Washington, DC: International Food Policy Research Institute, Discussion Paper 01054.

Raymond, C.M., I. Fazey, M.S. Reed, L.C. Stringer, G.M. Robinson and A.C. Evely. 2010. Integrating local and scientific knowledge for environmental management. *Journal of Environmental Management*, 91(8), 1,766–1,777.

Redford, K.H., W. Adams and G.M. Mace. 2013. Synthetic biology and conservation of nature: wicked problems and wicked solutions. *PLOS Biology*, 11(4), e1001530. Available at http://journals.plos.org/plosbiology/article?id=10.1371/journal.pbio.1001530. Accessed 15 September 2014.

Redmond, E.C. and C.J. Griffith. 2003. Consumer food handling in the home: a review of food safety studies. *Journal of Food Protection*, 66(1), 130–161.

REN21. 2014. Renewables 2014. *Global Status Report: Renewable Energy Policy Network for the 21st Century.* Paris, France: UNEP.

Rights of Mother Earth. 2014. *An Initial Report on Seeds & Seed Practices in the United States.* US Food Sovereignty Alliance, Rights of Mother Earth/Defense of the Commons Workgroup, April. Available at http://usfoodsovereigntyalliance.org/wp-content/uploads/2014/04/USFSASeedReportApril2014FINAL2small.pdf. Accessed 5 February 2016.

Rillig, M.C., B.A. Caldwell, H.A.B. Wösten and P. Sollins. 2007. Role of proteins in soil carbon and nitrogen storage: controls on persistence. *Biogeochemistry*, 85(1), 25–44.

Rip, A. and R. Kemp. 1998. Technological change. In *Human Choice and Climate Change*, Vol. 2, edited by S. Rayner and E.L. Malone. Columbus, OH: Battelle Press, 327–399.

Ritchie, K. 2012. *From Farm to Table: An Energy Consumption Assessment of Refrigerated, Frozen and Canned Food Delivery Scientific Certification Systems.* Available at http://www.steelforpackaging.org/uploads/ModuleXtender/Themesslides/6/3_2_1-FromFarmToTable.pdf. Accessed 5 February 2016.

Rocky Mountain Forest and Range Experiment Stations. 1963. *U.S. Forest Service Research Note.* Fort Collins, CO: Pacific Southwest Research Station.

Ruggeri, B. 2007. L'acqua e le sue implicazioni nel contesto sociale. Unpublished paper presented at 'L'acqua patrimonio dell'umanità' conference, 19 April, Padua, Italy.

Runyon, J. 2014. Spurred by Japan, steady growth predicted for energy storage market. *Renewable Energy World.com*, 19 March. Available at http://www.renewableenergyworld.com/articles/2014/03/spurred-by-japan-steady-growth-predicted-energy-storage-market.html. Accessed 13 September 2015.

Saleh, A.I. 2012. *Sociology of the Marsh Arabs.* London: AMAR International Foundation, Paper III.

Sarkis, J. 2006. The adoption of environmental and risk management practices: relationships to environmental performance. *Annals of Operations Research*, 145(1), 367–381.

Sassatelli, R. 2015. Consumer culture, sustainability and a new vision of consumer sovereignty. *Sociologia Ruralis*, 55(4), 483–496.

Sassatelli, R. and F. Davolio. 2010. Consumption, pleasure and politics: slow food and the politico-aesthetic problematization of food. *Journal of Consumer Culture*, 10(2), 202–232.

Scarborough, V.L. 2003. *The Flow of Power. Ancient Water Systems and Landscapes.* Santa Fe, NM: SAR Press.

Schirpke, U., G. Leitinger, E. Tasser, M. Schermer, M. Steinbacher and U. Tappeiner. 2013. Multiple ecosystem services of a changing Alpine landscape: past, present and future. *International Journal of Biodiversity Science, Ecosystem Services and Management*, 9(2): 123–135.

Schmidt, M. 2010. Xenobiology: a new form of life as the ultimate biosafety tool. *Bioessays*, 32(4), 322–331.

Schoenung, S.M. and W.V. Hassenzahl. 2003. *Long- vs. Short-term Energy Storage, Technologies Analysis: A Life-cycle Cost Study.* Albuquerque, NM: Sandia National Laboratories.

Scholte, J.A. 2004. Civil Society and democratically accountable global governance. *Government and Opposition*, 39(2), 211–233.

Selznick, P. 1957. *Leadership in Administration.* New York: Harper and Row.

Sen, A. 1991. Food, Economics, and Entitlements. In *The Political Economy of Hunger*, Vol. 1, edited by J. Drèze and A. Sen. Oxford: Oxford University Press, 34–52.

Sending, O.J. and I.B. Neumann. 2006. Governance to governmentality: analyzing NGOs, states, and power. *International Studies Quarterly*, 50(3), 651–672.

Sereni, E. 2014. *History of the Italian Agricultural Landscape.* Princeton, NJ: Princeton University Press; originally published 1961.

Sgrulletta, D., A, Cammerata, E. De Stefanis and L. Gazza. 2013. La rete qualità cereali italiana. *Dal Seme*, VIII(3), 62–65.

Shandra, J.M., C. Leckband, L.A. McKinney and B. London. 2009. Ecologically unequal exchange, world polity, and biodiversity loss: a cross-national analysis of threatened mammals. *International Journal of Comparative Sociology*, 50(3–4), 285–310.

Sharma, A., V.V. Tyagi, C.R. Chen and D. Buddhi. 2009. Review on thermal energy storage with phase change materials and applications. *Renewable and Sustainable Energy Reviews*, 13(2), 318–345.

Shiva, V. and G. Bedi. 2002. *Sustainable Agriculture and Food Security: The Impact of Globalization.* London: Sage.

Shum, K.L. and C. Watanabe. 2007. Photovoltaic deployment strategy in Japan and the USA: an institutional appraisal. *Energy Policy*, 35, 1,186–1,195.

Smith, L.D. and A.M. Thomson. 1991. *The Role of Public and Private Agents in the Food and Agricultural Sectors of Developing Countries.* Rome, Italy: Food and Agriculture Organization.

Smith, M.E. 2009. V. Gordon Childe and the urban revolution: a historical perspective on a revolution in urban studies. *Town Planning Review*, 80(1), 3–29.

Sobal, J. and C.A. Bisogni. 2009. Constructing food choice decisions. *Annals of Behavioral Medicine*, 38(1), supplement, 37–46.

Spaargaren, G. 2011. Theories of practices: agency, technology, and culture. exploring the relevance of practice theories for the governance of sustainable consumption practices in the new world order. *Global Environmental Change* 21(3), 813–822.

Spaargaren, G.A., A. Mol and H. Bruynincks. 2006. Introduction: governing environmental flows in global modernity. In *Governing Environmental Flows: Global Challenges to Social Theory,* edited by G. Spaargaren, A.P.J. Mol and F.H. Buttel. Cambridge MA: MIT Press, 1–36.

Spehn, E.M., M. Liberman and C. Korner, eds. 2006. *Land Use Change and Mountain Biodiversity.* London: CRC Press.

Spronk, S. 2010. Water and sanitation utilities in the global South: re-centering the debate on 'efficiency'. *Review of Radical Political Economics*, 42(2), 156–174.

St. John, J. 2013. California sets terms of massive energy storage mandate. *Greentech Media*, 4 September. Available at http://www.greentechmedia.com/articles/read/california-sets-terms-of-massive-energy-storage-mandate. Accessed 5 February 2016.

St. John, J. 2014. In Texas, a showdown over how to pay for grid batteries. *Greentech Media*, 3 December. Available at http://www.greentechmedia.com/articles/read/in-texas-a-conflict-over-grid-batteries. Accessed 5 February 2016.

Steffen, W., A.A. Burbidge, L. Hughes, R. Kitching, D. Lindenmayer, W. Musgrave, M. Stafford Smith and P.A. Werner, eds. 2009. *Australia's Biodiversity and Climate Change*, Melbourne, Australia: CSIRO Publishing.

Steinhour, S. 2013. *Reaching Agreement on Minimizing Damage by Extreme Weather is Easier if We Plan for Uncertainty.* Mill Valley, CA: Resource Renewal Institute.

Strassburg, B.B.N., A. Kelly, A. Balmford, R.G. Davies, H.K. Gibbs, A. Lovett, L. Miles, C.D.L. Orme, J. Price, R.K. Turner and A.S.L. Rodrigues. 2010. Global congruence of carbon storage and biodiversity in terrestrial ecosystems. *Conservation Letters*, 3(2), 98–105.

Stuiver, M. 2006. Highlighting the retro side of innovation and its potential for regime change in agriculture. *Research in Rural Sociology and Development*, 12, 147–173.

Sturgeon, T.J. 2002. *Modular Production Networks: New American Model of Industrial Organization.* Cambridge, MA: MIT Industrial Performance Center, MIT Working Paper IPC-02-003.

Subramaniam, M. and C. Bunka. 2013. Food security and state: policy considerations for the contemporary food crisis. *Global Policy Research Institute (GPRI) Policy Briefs*, 1(1), Article 7. Available at http://docs.lib.purdue.edu/gpripb/vol1/iss1/7. Accessed 5 February 2016.

Svarstad, H., L.K. Petersen, D. Rothman, H. Siepel and F. Wätzold. 2008. Discursive biases of the environmental research framework DPSIR. Land Use Policy, 25(1), 116–125.

Swyngedouw, E. 2005. Dispossessing H_2O: the contested terrain of water privatization. *Capitalism Nature Socialism*, 16(1), 81–98.

Tainter, J. 1990. *The Collapse of Complex Societies.* Cambridge: Cambridge University Press.

TERI-IGES, AEI. 2012. *Learning from Emerging Energy Innovations in Asia: Contributing to the Discourse on an Institutional Framework for Sustainable Development.* New Delhi, India: Energy and Resources Institute.

Terpstra, M.J., L.P.A. Steenbekkers, N.C.M. de Maertelaere and S. Nijhuis. 2005. Food storage and disposal: consumer practices and knowledge. *British Food Journal*, 107(7), 526–533.

Ticchiati, V. 2008. Stoccaggio. L'insostituibile ruolo dei commercianti. *AgriCommercio*, 5–6, 11–14.

Tomita, T. 2014. Policies and regulations for electricity storage in Japan. Paper presented at IRENA. International Energy Storage Policy and Regulation Workshop, Düsseldorf, Germany, 27 March.

Troy, A. and B. Voigt. 2012. *Analysis of the Drivers of Urban Growth and Second Home Development in the Northern Forest Region of Vermont: Project Summary.* Burlington, VT: University of Vermont.

Tscharntke, T., Y. Clough, T.C. Wanger, L. Jackson, I. Motzke, I. Perfecto, J. Vandermeer and A. Whitbread. 2012. Global food security, biodiversity conservation and the future of agricultural intensification. *Biological Conservation*, 151(1), 53–59.

Tucker, M.E. 2006. Religion and ecology: survey of the field. In *The Oxford Handbook of Religion and Ecology*, edited by R.S. Gottlieb and M.E. Tucker. Oxford: Oxford University Press.

Turnhout, E. and S. Boonman-Berson. 2011. Databases, scaling practices, and the globalization of biodiversity. *Ecology and Society*, 16(1), 35.

Tweed, K. 2013. Community energy storage: almost, but not quite there. *The Energy Collective*, 12 July. Available at http://www.theenergycollective.com/stephenlacey/248301/community-battery-storage-almost-not-quite-there. Accessed 5 February 2016.

Uhlaner, C.J. 1989, Relational goods and participation: incorporating sociability into a theory of rational action. *Public Choice*, 62(3), 253–285.

Ulu, E.Y., E. Cetin, O.O. Karakilinc, A. Yilanci and H.K. Ozturk. 2013. Analysis of a photovoltaic–fuel cell hybrid energy system in terms of electromagnetic pollution. *International Journal on Energy Conversion*, 1(1), 14–22.

United Nations Environment Programme. 2005. *Integrated Assessment of the Impact of Trade Liberalization: A Country Study on the Indonesian Rice Sector*. Geneva, Switzerland: UNEP/Earthprint.

Urry, J. 2010. Sociology facing climate change. *Sociological Research Online*, 15(3), 1.

Van der Ploeg, J. 2008. *The New Peasantries: Struggle of Autonomy and Sustainability in an Era of Empire and Globalisation*. London: Earthscan.

Van Koppen, K. 2006. Governing nature? On the global complexity of biodiversity conservation. In *Governing Environmental Flows: Global Challenges to Social Theory*, edited by G. Spaargaren, A.P.J. Mol and F.H. Buttel. Cambridge MA: MIT Press, 187–220.

Van Koppen, K. and W.T. Markham. 2007. Nature protection in Western environmentalism: a comparative analysis. In *Protecting Nature: Organizations and Networks in Europe and the USA*, edited by K. Van Koppen and W.T. Markham. Cheltenham: Elgar, 263–285.

Van Vliet, B., E. Shove and H. Chappells. 2012. *Infrastructures of Consumption: Environmental Innovation in the Utility Industries*. London: Earthscan.

Venolia, C. 2011. Aggressively passive: building homes to the passive house standard. *Mother Earth Living*, January/February. Available at http://www.motherearthliving.com/Green-Homes/aggressively-passive-building-homes-to-the-passive-house-standard.aspx. Accessed 21 September 2015.

Verda, V. and F. Colella. 2011. Primary energy savings through thermal storage in district heating networks. *Energy*, 36(7), 4,278–4,286.

Verhoeven, J.T.A. and T.L. Setter. 2010. Agricultural use of wetlands: opportunities and limitations. *Annals of Botany*, 105(1), 155–163.

Vernooy, R. 2013. In the hands of many: a review of community gene/seed banks around the world. In *Community Seed Banks in Nepal: Past, Present, Future*, edited by P. Shrestha, R. Vernooy and P. Chaudhary. Pokhara, Nepal: Local Initiatives for Biodiversity, Research and Development, 3–15.

Veronese, G. 1925. *L'epopea delle bonifiche private*. Federazione nazionale delle bonifiche. Padua, Italy: Soc. cooperativa tipografica.

Verweij, M. 2001. Towards sustainable pond farming. *LEISA Magazine*, 17(3), 43–45.

Verweij, M. and M. Thompson, eds. 2006. *Clumsy Solutions for a Complex World: Governance, Politics and Plural, Perceptions*. Basingstoke: Palgrave Macmillan.

Viale, R. and A. Pozzali. 2010. Complex adaptive systems and the evolutionary triple helix. *Critical Sociology*, 36(4), 575–594.
Von Ammon, M. and D. Pohl. 2013. *Der deutsche Markt für Solarstromspeicher – Erfahrungen und Einschätzungen*. Bonn, Germany: EuPD Research.
Wang, S. 2006. Money and autonomy: patterns of civil society finance and their implications. *Studies in Comparative International Development*, 40(4), 3–29.
Wang, U. 2013. 5 reasons you should care about California's new energy storage mandate. *GIGAOM*, 17 October. Available at http://gigaom.com/2013/10/17/5-reasons-you-should-care-about-californias-new-energy-storage-mandate/. Accessed 8 August 2014.
Ward, L. 2013. Eco-governmentality revisited: mapping divergent subjectivities among integrated water resource management experts in Paraguay. *Geoforum*, 46, 91–102.
Wassmann, R., N.X. Hien, C.T. Hoanh and T.P. Tuong. 2004. Sea level rise affecting the Vietnamese Mekong Delta: water elevation in the flood season and implications for rice production. *Climatic Change*, 66, 89–107.
Watanabe, C., K. Wakabayashi and T. Miyazawa. 2000. Industrial dynamism and the creation of a 'virtuous cycle' between R&D, market growth and price reduction: the case of photovoltaic power generation (PV) development in Japan. *Technovation*, 20(6), 299–312.
Waughray, D., ed. 2011. *Water Security: The Water–Food–Energy–Climate Nexus*. Washington, DC: Island Press.
Weber, M. 1905. *Die protestantische Ethik und der Geist des Kapitalismus*, I and II. In *Archiv für Sozialwissenschaft und Sozialpolitik*, XX, 1–54, and XXI, 1–110.
Weber, M. 1922. *Wirtschaft und Gesellschaft*. Tübingen, Germany: Mohr.
Weber, O. 2012. Sustainable banking – history and current developments. *EMES-SOCENT Conference Selected Papers*, ECSP-LG13-39. Brussels, Belgium: Belgian Science Policy Office.
Wenger E. 1998. Communities of practice and social learning systems. *Organization*, 7(2), 225–246.
Wesoff, E. 2014. California approves $415m for behind-the-meter energy storage, fuel cells, wind and turbines. *Greentech Media*, 18 June. Available at http://www.greentechmedia.com/articles/read/California-Approves-415M-For-Behind-the-Meter-Storage-Fuel-Cells-Wind-an. Accessed 5 February 2015.
Wirth, H. 2014. *Recent Facts about Photovoltaics in Germany*. Freiburg, Germany: Fraunhofer Institute for Solar Energy Systems.
Wittfogel, K. 1957. *Oriental Despotism: A Comparative Study of Total Power*. New Haven, CT: Yale University Press.
Wolfe, C. 2010. Language. In *Critical Terms for Media Studies*, edited by M.B.N. Mitchell and W.J.T Hansen. Chicago, IL: University of Chicago Press.
Wood, E. 2014. Microgrids and utilities: friend or foe? *Microgrid Knowledge*, 18 June. Available at http://www.energyefficiencymarkets.com/microgrids-utilities-friends-foe/. Accessed 7 August 2014.
Woodburn, J. 1982. Egalitarian societies. *Man*, 17(3), 431–51.
World Packaging Organisation. 2008. *Market Statistics and Future Trends in Global Packaging*. Available at http://www.worldpackaging.org/i4a/doclibrary/getfile.cfm?doc_id=7. Accessed 5 February 2016.
World Trade Organization. 2013. *World Trade Report 2013: Factors Shaping the Future of World Trade*. Geneva, Switzerland: WTO.
Wundt, W.M. 1897. *Ethics: An Investigation of the Facts and Laws of the Moral Life*. Translated from the 2nd German edn, 1892. London: S. Sonnenschein.

Wynne, B. 2013. *Rationality and Ritual: Participation and Exclusion in Nuclear Decision-making*. London: Routledge.

Young, C. and E. Quinn. 2015. Food safety scientists have ties to Big Tobacco. *The Center for Public Integrity*, 15 April. Available at http://www.publicintegrity.org/2015/04/15/17144/food-safety-scientists-have-ties-big-tobacco. Accessed 5 February 2016.

Zach, K., H. Auer and G. Lettner. 2012. *Facilitating Energy Storage to Allow High Penetration of Intermittent Renewable Energy*. Vienna, Austria: Intelligent Energy Europe.

Zahrnt, V. 2011. *Food Security and the EU's Common Agricultural Policy: Facts against Fears*. Brussels, Belgium: ECIPE Working Paper no. 01/2011.

Ziv, G., E. Baran, S. Nam, I. Rodríguez-Iturbe and S.A. Levin. 2012. Trading-off fish biodiversity, food security, and hydropower in the Mekong River Basin. *Proceedings of National Academy of Science of United States of America*, 109(15): 5,609–5,614.

Zwarteveen, M.Z. and R. Boelens. 2014. Defining, researching and struggling for water justice: some conceptual building blocks for research and action. *Water International*, 39(2), 143–158.

Index

accessibility 22–3, 79, 81, 97
advocacy task 92
alignment 124–5
anthropocene 152
article of trade 2; *see also* commodity
association (business/professional) 24, 30–2, 42, 44, 92, 103, 118, 122, 132, 149, 151, 169, 171

behind-the-meter 121, 124–5, 136
boundary: object 93; work 24
break/exempt/relief (tax) 73, 102, 114, 146
buffer 5, 91, 156, 164

carbon capture 15, 158–9
carbon storage pools 158
chain: cold 41, 43; energy 100,104, 117–18, 122; food 24, 37, 40, 44, 56
civicness 87, 89
civil protection xii, 25, 55, 169, 171
civilization xiv, 11–13, 16, 58, 141
climate change xiv, 9, 18, 53, 66, 68, 89, 158–9, 168, 172, 175
combined heat and power-CHP 105, 121, 132–3
commodity xii, 5–6, 9, 95
commons 4–5, 44, 69, 79, 152
community seed bank 149, 151
concentrated solar 105, 115, 126, 128–9, 133
consumerism 9, 13, 22
controlled/programmed floods 68–76
cooperation 33–4, 83, 94, 121
cooperative 29–33, 134
coordinated economy 121, 126

cosmology xiii, 47, 56
customization 94, 125

dependency (past/path) 101, 108, 130
diet xii, 10, 35, 45
differentiation 13, 18–19, 27, 30–1, 33–4, 39, 41, 169, 173
distance 2, 8, 17, 52, 77–8, 94
district heating 127–8, 130–5, 174

ecological footprint 3
embeddedness 15, 104
exchange: enlarged 95–6; food/seed 145–6, 152, 169; gift 14, 17, 170–1; information/knowledge 44, 48, 96; market 94; stock 5, 83; terms of 110, 147

farmer 4, 11–12, 26, 29–33, 61, 63, 65, 68, 72, 74–6, 88, 90–2, 145, 149, 151, 169–70, 173
farming 61–2, 92, 147, 154, 158
floodplain 66–8, 70, 74–5
forestry/agro-forestry 68, 155, 157–8

global food 26–7
governmentality 21–3, 87, 95
gradualism 135, 143, 156
green economy 57, 74, 102, 112

hegemony 62–5
heterogenesis (ends/goals) 45, 149

identity 13–14, 51, 140, 142, 171
ideology 48, 62, 81

Index 201

impoundment 90, 166
incentive 30, 64, 92, 102–4, 114–15, 121–2, 130–1, 134, 159, 173; *see also* subsidies
institutional 15, 18, 104, 121, 125, 134, 136, 138, 143, 146, 150
intellectual property 151–2
intervention *see* policy
irrigation 12, 58–62, 83, 88–9, 92–3, 140, 154

knowledge: ethnic/local/traditional 16, 74, 143; expert/scientific 35, 37, 74, 150

landowner xiii, 4, 61, 72, 75–6, 142, 172,
landscape 13, 16, 59–60, 83, 88–9, 91, 101, 104, 139–42, 154–7, 165, 175
learning xiv, 11, 71, 73, 117, 140, 152, 170
legitimization/legitimacy/legitimate 15, 38, 57, 81, 149
lifecycle 106, 109
lifestyle xii, 10, 23, 55, 69, 112

matrix 165
mobility 16–17, 43, 62, 163
modularity 2, 21 27, 47, 135, 168–9
modularization 36, 40, 56
monopoly 77, 94, 111–12, 118–19, 123, 125, 169
movement (social) xiii, 10, 23, 36, 48, 51–3, 58, 61, 63, 82–3, 92, 117, 119–20, 124, 139–40

nature park 63, 141–3, 150, 152, 170
net metering 111, 131, 134–5
network (social) 16, 31, 175
NGOs 93, 95, 143, 150, 170
non-profit 40, 46, 92, 118, 146–51, 164

organization (environmental) 35, 139, 149

palafitte 69, 74
participation 57, 74–5, 92, 95, 143, 163
passive house 43, 69, 169
patent/patented 127, 150, 164, 174
peasant 152, 174

policy: Common Agriculture 28–9, 173; energy 115, 134–5; risk assessment 37–8; storage 29–30, 122–3, 125, 172–5
political economy 21, 114, 123
precautionary principle 38–9
preparedness xiii, xiv, 15–16, 47–50, 52, 54–6, 70, 75–6
preppers 50, 53–5
preservation (food) 19–20, 24, 26, 33–9, 41–5, 52–7
primate city 154
private property 89–90, 152
privatization 82, 93, 95
prosumer 105, 111
protected areas 63, 68, 71, 138, 142–3, 148–52, 156–7, 164–5, 174–5; *see also* nature park
pumped hydro (power) 99–100, 102–3, 106, 116, 126

rational choice 7, 14
rationalisation 6, 34, 57, 61–2, 65, 83, 170
rationality xiv, 4, 24–5, 34, 47–9, 57, 76, 87, 171
rationing 56–7, 87, 172
rebound effect 81
reciprocity 16–18, 171
reclamation (land) xiii, xv, 32, 58–65, 86
reduce emissions from deforestation and degradation-REDD 159
re-entry 157
reflexivity 11
regime 3, 26–7, 60, 104, 118–19, 131
relational 13–14, 95, 112; *see also* reciprocity
renewables 100–36, 172
retro-innovation xiv, 14, 44
rising expectations 81
risk: analysis 16, 36; management 15–16, 37, 71; reduction 17, 54

sanctuary (animal) 138, 140, 153
self reliance 47–8, 87, 169
shadow price 42
smart grid 96, 110, 131
social bank 152
social formation 21, 62

socio-technical xii, xv, 13, 19, 59, 76, 97, 107, 112, 129, 135
source/sink 164–5
sovereignty 34, 61
storage: economy 50; fix 15; *see also* policy
storing by moving 25–6
subsidiarity 87, 131, 156
subsidies 29–30, 64, 109–10, 113–15, 121–5, 130, 136, 172–3; *see also* incentives
survivalism 25, 52–3, 55
symbolization 23–4, 49–52

tax/taxation xi, 30, 68, 72, 79, 94, 172; *see also* break/exempt/relief
technological: dense environment/system 108, 133; fix 16; neutrality 122, 125, 136, 172; package 87, 93, 102, 104, 115, 135; path 36, 44, 60, 87, 101–3, 112, 130

third sector 57, 92
trade/commerce xi, 7–9, 25–32, 44, 124, 148, 152
trade-offs 68, 111, 161, 164
trajectory 15–16, 104, 106–8, 112, 132
transition 15, 99, 104, 112, 114, 119–21, 134
transmission system operator-TSO 117–19, 123, 129
transparency 39, 94, 115
triple helix 15, 38, 170

unbundling 100, 112, 119, 123–4
universal 39, 52, 93, 148–9, 151
utility 82–4, 87, 93–6, 108–26, 135–7, 171

voluntary xiv, 74, 84, 87, 134
volunteers 47, 55, 169

For Product Safety Concerns and Information please contact our
EU representative GPSR@taylorandfrancis.com Taylor & Francis
Verlag GmbH, Kaufingerstraße 24, 80331 München, Germany